Succeed

Eureka Math®
Grade 3
Modules 1 & 2

TEKS EDITION

Great Minds® is the creator of *Eureka Math*®, *Wit & Wisdom*®, *Alexandria Plan*™, and *PhD Science*®.

Published by Great Minds PBC
greatminds.org

© 2020 Great Minds PBC. Except where otherwise noted, this content is published under a limited license with the Texas Education Agency. Use is limited to noncommercial educational purposes. Where indicated, teachers may copy pages for use by students in their classrooms. For more information, visit http://gm.greatminds.org/texas.

Printed in the USA

1 2 3 4 5 6 7 8 9 10 CCR 25 24 23 22

ISBN 978-1-63642-863-5

Learn ♦ Practice ♦ Succeed

Eureka Math® student materials for *A Story of Units®* (K–5) are available in the *Learn, Practice, Succeed* trio. This series supports differentiation and remediation while keeping student materials organized and accessible. Educators will find that the *Learn, Practice,* and *Succeed* series also offers coherent—and therefore, more effective—resources for Response to Intervention (RTI), extra practice, and summer learning.

Learn

Eureka Math Learn serves as a student's in-class companion where they show their thinking, share what they know, and watch their knowledge build every day. *Learn* assembles the daily classwork—Application Problems, Exit Tickets, Problem Sets, templates—in an easily stored and navigated volume.

Practice

Each *Eureka Math* lesson begins with a series of energetic, joyous fluency activities, including those found in *Eureka Math Practice*. Students who are fluent in their math facts can master more material more deeply. With *Practice,* students build competence in newly acquired skills and reinforce previous learning in preparation for the next lesson.

Together, *Learn* and *Practice* provide all the print materials students will use for their core math instruction.

Succeed

Eureka Math Succeed enables students to work individually toward mastery. These additional problem sets align lesson by lesson with classroom instruction, making them ideal for use as homework or extra practice. Each problem set is accompanied by a Homework Helper, a set of worked examples that illustrate how to solve similar problems.

Teachers and tutors can use *Succeed* books from prior grade levels as curriculum-consistent tools for filling gaps in foundational knowledge. Students will thrive and progress more quickly as familiar models facilitate connections to their current grade-level content.

Students, families, and educators:

Thank you for being part of the *Eureka Math®* community, where we celebrate the joy, wonder, and thrill of mathematics.

Nothing beats the satisfaction of success—the more competent students become, the greater their motivation and engagement. The *Eureka Math Succeed* book provides the guidance and extra practice students need to shore up foundational knowledge and build mastery with new material.

What is in the Succeed *book?*

Eureka Math Succeed books deliver supported practice sets that parallel the lessons of *A Story of Units®*. Each *Succeed* lesson begins with a set of worked examples, called *Homework Helpers*, that illustrate the modeling and reasoning the curriculum uses to build understanding. Next, students receive scaffolded practice through a series of problems carefully sequenced to begin from a place of confidence and add incremental complexity.

How should Succeed *be used?*

The collection of *Succeed* books can be used as differentiated instruction, practice, homework, or intervention. When coupled with *Affirm®*, *Eureka Math*'s digital assessment system, *Succeed* lessons enable educators to give targeted practice and to assess student progress. *Succeed*'s perfect alignment with the mathematical models and language used across *A Story of Units* ensures that students feel the connections and relevance to their daily instruction, whether they are working on foundational skills or getting extra practice on the current topic.

Where can I learn more about Eureka Math *resources?*

The Great Minds® team is committed to supporting students, families, and educators with an ever-growing library of resources, available at gm.greatminds.org/math-for-texas. The website also offers inspiring stories of success in the *Eureka Math* community. Share your insights and accomplishments with fellow users by becoming a *Eureka Math* Champion.

Best wishes for a year filled with Eureka moments!

Jill Diniz
Director of Mathematics
Great Minds

Contents

Module 1: Properties of Multiplication and Division and Solving Problems with Units of 2–5 and 10

Topic A: Multiplication and the Meaning of the Factors
Lesson 1 3
Lesson 2 7
Lesson 3 11

Topic B: Division as an Unknown Factor Problem
Lesson 4 15
Lesson 5 19
Lesson 6 23

Topic C: Multiplication Using Units of 2 and 3
Lesson 7 27
Lesson 8 31
Lesson 9 35
Lesson 10 39

Topic D: Division Using Units of 2 and 3
Lesson 11 43
Lesson 12 47
Lesson 13 51

Topic E: Multiplication and Division Using Units of 4
Lesson 14 55
Lesson 15 59
Lesson 16 63
Lesson 17 67

Topic F: Distributive Property and Problem Solving Using Units of 2–5 and 10
Lesson 18 71
Lesson 19 75
Lesson 20 79
Lesson 21 83

Module 2: Place Value and Problem Solving with Units of Measure

Topic A: Time Measurement and Problem Solving

Lesson 1 89

Lesson 2 93

Lesson 3 97

Topic B: Measuring Weight and Liquid Volume in Metric Units

Lesson 4 101

Lesson 5 105

Lesson 6 109

Lesson 7 113

Lesson 8 117

Lesson 9 121

Topic C: Place Value and Comparing Multi-Digit Whole Numbers

Lesson 10 125

Lesson 11 131

Lesson 12 135

Topic D: Rounding to the Nearest Ten, Hundred, Thousand, and Ten Thousand

Lesson 13 141

Lesson 14 145

Lesson 15 149

Lesson 16 153

Topic E: Two- and Three-Digit Measurement Addition Using the Standard Algorith

Lesson 17 157

Lesson 18 161

Lesson 19 165

Topic F: Two- and Three-Digit Measurement Subtraction Using the Standard Algorithm

Lesson 20 169

Lesson 21 173

Lesson 22 177

Lesson 23 181

Grade 3
Module 1

1. Solve each number sentence.

> I know this picture shows equal groups because each group has the same number of triangles. There are 3 equal groups of 4 triangles.

3 groups of 4 = **12**

3 fours = **12**

4 + 4 + 4 = **12**
3 × 4 = **12**

> I can multiply to find the total number of triangles because multiplication is the same as repeated addition! 3 groups of 4 is the same as 3 × 4. There are 12 total triangles, so 3 × 4 = 12.

2. Circle the picture that shows 3 × 2.

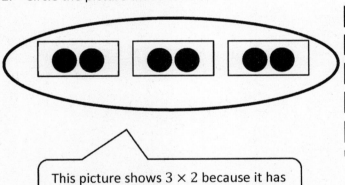

> This picture shows 3 × 2 because it has 3 groups of 2. The groups are equal.

> This picture does *not* show 3 × 2 because the groups are not equal. Two of the groups contain 2 objects, but the other only has 1 object.

Lesson 1: Understand *equal groups of* as multiplication.

Name _____ Date _____

1. Fill in the blanks to make true statements.

a. 4 groups of five = _____

 4 fives = _____

 4 × 5 = _____

b. 5 groups of four = _____

 5 fours = _____

 5 × 4 = _____

c. 6 + 6 + 6 = _____

 _____ groups of six = _____

 3 × _____ = _____

d. 3 + ____ + ____ + ____ + ____ + ____ = ____

 6 groups of _____ = _____

 6 × _____ = _____

Lesson 1: Understand *equal groups of* as multiplication.

2. The picture below shows 3 groups of hot dogs. Does the picture show 3 × 3? Explain why or why not.

3. Draw a picture to show 4 × 2 = 8.

4. Circle the pencils below to show 3 groups of 6. Write a repeated addition and a multiplication sentence to represent the picture.

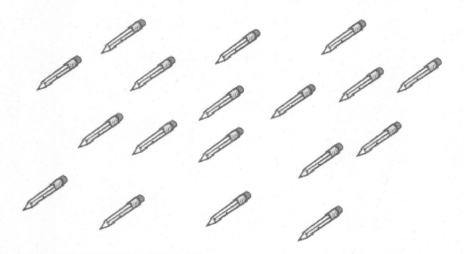

Lesson 1: Understand *equal groups of* as multiplication.

A STORY OF UNITS – TEKS EDITION

Lesson 2 Homework Helper 3•1

1. Use the array below to answer the questions.

> The hearts are arranged in an array, and I know that a row in an array goes straight across. There are 5 rows in this array. Each row has 4 hearts.

a. What is the number of rows? __5__

b. What is the number of objects in each row? __4__

c. Write a multiplication expression to describe the array. __5 × 4__

> I know a multiplication expression is different from an equation because it doesn't have an equal sign.

> I can write the expression 5 × 4 because there are 5 rows with 4 hearts in each row.

2. The triangles below show 2 groups of four.

a. Redraw the triangles as an array that shows 2 rows of four.

> I can redraw the equal groups as an array. I can draw 2 rows with 4 triangles in each row.

b. Compare the groups of triangles to your array. How are they the same? How are they different?

They are the same because they both have the same number of triangles, 8. They are different because the triangles in the array are in rows, but the other triangles are not in rows.

> I need to make sure to explain how they are the same *and* how they are different!

Lesson 2: Relate multiplication to the array model.

3. Kimberly arranges her 14 markers as an array. Draw an array that Kimberly might make. Then, write a multiplication equation to describe your array.

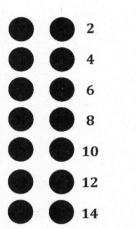

$7 \times 2 = 14$

I can write the equation by writing the number of rows (groups), 7, times the number in each group, 2. The product (total) is 14.

This problem doesn't tell me the number of rows or the number of objects in each row. I need to use the total, 14, to make an array. Since 14 is an even number, I am going to make rows of 2. I can skip count by 2 and stop when I get to 14.

I think there are other arrays that would work for a total of 14. I can't wait to see what my friends came up with!

Lesson 2: Relate multiplication to the array model.

Name _____ Date _____

Use the arrays below to answer each set of questions.

1.   a. How many rows of erasers are there? _____

   b. How many erasers are there in each row? _____

2.    a. What is the number of rows? _____

   b. What is the number of objects in each row? _____

3. 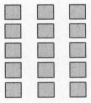   a. There are 3 squares in each row. How many squares are in 5 rows? _____

   b. Write a multiplication expression to describe the array. _____

4.    a. There are 6 rows of stars. How many stars are in each row? _____

   b. Write a multiplication expression to describe the array. _____

Lesson 2:   Relate multiplication to the array model.

5. The triangles below show 3 groups of four.

    a. Redraw the triangles as an array that shows 3 rows of four.

    b. Compare the drawing to your array. How are they the same? How are they different?

6. Roger has a collection of stamps. He arranges the stamps into 5 rows of four. Draw an array to represent Roger's stamps. Then, write a multiplication equation to describe the array.

7. Kimberly arranges her 18 markers as an array. Draw an array that Kimberly might make. Then, write a multiplication equation to describe your array.

1. There are ___3___ apples in each basket. How many apples are there in 6 baskets?

   a. Number of groups: ___6___   Size of each group: ___3___

   b. 6 × ___3___ = ___18___

   c. There are ___18___ apples altogether.

> Each circle represents 1 basket of apples. There are 6 circles with 3 apples in each circle. The number of groups is 6, and the size of each group is 3. There are 18 apples altogether. I can show this with the equation $6 \times 3 = 18$.

2. There are 3 bananas in each row. How many bananas are there in ___4___ rows?

   a. Number of rows: ___4___   Size of each row: ___3___

   b. ___4___ × 3 = ___12___

   c. There are ___12___ bananas altogether.

> I can show this with the equation $4 \times 3 = 12$. The 4 in the equation is the number of rows, and 3 is the size of each row.

Lesson 3: Interpret the meaning of factors—the size of the group or the number of groups.

A STORY OF UNITS – TEKS EDITION
Lesson 3 Homework Helper 3•1

> The factors tell me the number of groups and the size of each group. I can draw an array with 3 rows and 5 in each row.

3. Draw an array using factors 3 and 5. Then, show a number bond where each part represents the amount in one row.

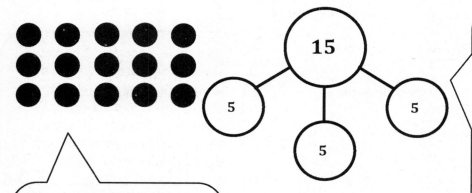

> My array shows 3 rows of 5. I could have used the same factors, 3 and 5, to draw an array with 5 rows of 3. Then my number bond would have 5 parts, and each part would have a value of 3.

> A number bond shows a part–whole relationship. I can draw a number bond with a total of 15 because there are 15 dots in my array. I can draw 3 parts for my number bond because there are 3 rows in my array. I can label each part in my number bond as 5 because the size of each row is 5.

Lesson 3: Interpret the meaning of factors—the size of the group or the number of groups.

Name _____ Date _____

Solve Problems 1–4 using the pictures provided for each problem.

1. There are 5 pineapples in each group. How many pineapples are there in 5 groups?

   a. Number of groups: _____   Size of each group: _____

   b. 5 × 5 = _____

   c. There are _____ pineapples altogether.

2. There are _____ apples in each basket. How many apples are there in 6 baskets?

   a. Number of groups: _____   Size of each group: _____

   b. 6 × _____ = _____

   c. There are _____ apples altogether.

3. There are 4 bananas in each row. How many bananas are there in _____ rows?

   a. Number of rows: _____   Size of each row: _____

   b. _____ × 4 = _____

   c. There are _____ bananas altogether.

4. There are _____ peppers in each row. How many peppers are there in 6 rows?

   a. Number of rows: _____   Size of each row: _____

   b. _____ × _____ = _____

   c. There are _____ peppers altogether.

5. Draw an array using factors 4 and 2. Then, show a number bond where each part represents the amount in one row.

Lesson 3: Interpret the meaning of factors—the size of the group or the number of groups.

1. Fill in the blanks.

   > The chickens are arranged in an array. I know there are 12 chickens divided equally into 3 groups since each row represents 1 equal group. Each group (row) has 4 chickens. So, the answer in my division sentence, 4, represents the size of the group.

   __12__ chickens are divided into __3__ equal groups.

   There are __4__ chickens in each group.

   $12 \div 3 =$ __4__

2. Grace has 16 markers. The picture shows how she placed them on her table. Write a division sentence to represent how she equally grouped her markers.

   There are __4__ markers in each row.

   __16__ $\div$ __4__ $=$ __4__

   > I can write the total number of markers Grace has, 16, since a division equation begins with the total.

   > The 4 represents the number of equal groups. I know there are 4 equal groups because the array shows 4 rows of markers.

   > This 4 represents the size of the group. I know this because the array shows 4 markers in each row.

Lesson 4: Understand the meaning of the unknown as the size of the group in division.

Name _____ Date _____

1.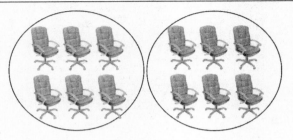

   12 chairs are divided into 2 equal groups.

   There are _____ chairs in each group.

2.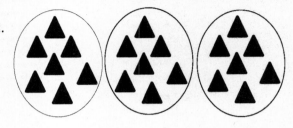

   21 triangles are divided into 3 equal groups.

   There are _____ triangles in each group.

3.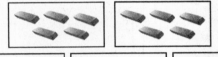

   25 erasers are divided into _____ equal groups.

   There are _____ erasers in each group.

4.

   _____ chickens are divided into _____ equal groups.

   There are _____ chickens in each group.

   9 ÷ 3 = _____

5.

   There are _____ buckets in each group.

   12 ÷ 4 = _____

6.

   16 ÷ 4 = _____

Lesson 4: Understand the meaning of the unknown as the size of the group in division.

7. Andrew has 21 keys. He puts them in 3 equal groups. How many keys are in each group?

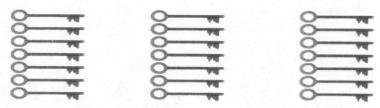

There are _____ keys in each group.

21 ÷ 3 = _____

8. Mr. Doyle has 20 pencils. He divides them equally between 4 tables. Draw the pencils on each table.

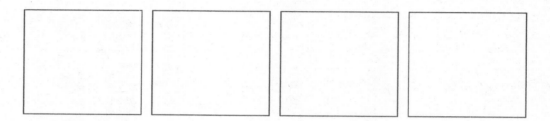

There are _____ pencils on each table.

20 ÷ _____ = _____

9. Jenna has markers. The picture shows how she placed them on her desk. Write a division sentence to represent how she equally grouped her markers.

There are _____ markers in each row.

_____ ÷ _____ = _____

Lesson 4: Understand the meaning of the unknown as the size of the group in division.

A STORY OF UNITS – TEKS EDITION

Lesson 5 Homework Helper 3•1

1. Group the squares to show $8 \div 4 =$ _____ where the unknown represents the number of groups.

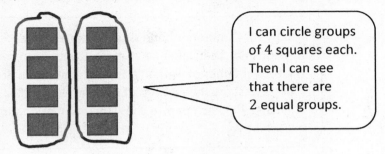

I can circle groups of 4 squares each. Then I can see that there are 2 equal groups.

How many groups are there? __2__

$8 \div 4 =$ __2__

2. Nathan has 14 apples. He puts 7 apples in each basket. Circle the apples to find the number of baskets Nathan fills.

I can circle groups of 7 apples to find the total number of baskets Nathan fills, 2 baskets.

a. Write a division sentence where the answer represents the number of baskets that Nathan fills.

__14__ ÷ __7__ = __2__

I can write a division sentence beginning with the total number of apples, 14, divided by the number of apples in each basket, 7, to find the number of Nathan's baskets, 2. I can check my answer by comparing it to the circled picture above.

Lesson 5: Understand the meaning of the unknown as the number of groups in division.

b. Draw a number bond to represent the problem.

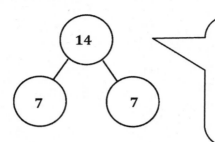

I know that a number bond shows a part–whole relationship. I can label 14 as my whole to represent the total number of Nathan's apples. Then I can draw 2 parts to show the number of baskets Nathan fills and label 7 in each part to show the number of apples in each basket.

3. Lily draws tables. She draws 4 legs on each table for a total of 20 legs.

   a. Use a count-by to find the number of tables Lily draws. Make a drawing to match your counting.

   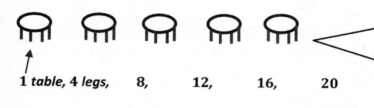

   I can draw models to represent each of Lily's tables. As I draw each table, I can count by four until I reach 20. Then, I can count to find the number of tables Lily draws, 5 tables.

   b. Write a division sentence to represent the problem.

   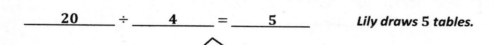

   _____20_____ ÷ _____4_____ = _____5_____   Lily draws 5 tables.

   I can write a division sentence beginning with the total number of legs, 20, divided by the number of legs on each table, 4, to find the number of tables Lily draws, 5. I can check my answer by comparing it to my picture and count-by in part (a).

Lesson 5: Understand the meaning of the unknown as the number of groups in division.

Name _____ Date _____

1.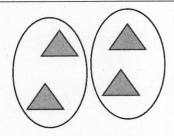

   Divide 4 triangles into groups of 2.

   There are _____ groups of 2 triangles.

   4 ÷ 2 = 2

2.

   Divide 9 eggs into groups of 3.

   There are _____ groups.

   9 ÷ 3 = _____

3.

   Divide 12 buckets of paint into groups of 3.

   12 ÷ 3 = _____

4.

   Group the squares to show 15 ÷ 5 = _____, where the unknown represents the number of groups.

   How many groups are there? _____

Lesson 5: Understand the meaning of the unknown as the number of groups in division.

5. Daniel has 12 apples. He puts 6 apples in each bag. Circle the apples to find the number of bags Daniel makes.

a. Write a division sentence where the answer represents the number of Daniel's bags.

b. Draw a number bond to represent the problem.

6. Jacob draws cats. He draws 4 legs on each cat for a total of 24 legs.

a. Use a count-by to find the number of cats Jacob draws. Make a drawing to match your counting.

b. Write a division sentence to represent the problem.

1. Sharon washes 20 bowls. She then dries and stacks the bowls equally into 5 piles. How many bowls are in each pile?

   $20 \div 5 = $ __4__

   $5 \times $ __4__ $= 20$

   I can draw an array with 5 rows to represent Sharon's piles of bowls. I can keep drawing columns of 5 dots until I have a total of 20 dots. The number in each row shows how many bowls are in each pile.

   What is the meaning of the unknown factor and quotient? __*It represents the size of the group.*__

   I know that the quotient is the answer you get when you divide one number by another number.

   I can see from my array that both the unknown factor and quotient represent the size of the group.

2. John solves the equation _____ $\times 5 = 35$ by writing and solving $35 \div 5 =$ _____. Explain why John's method works.

   John's method works because in both problems there are 7 groups of 5 and a total of 35. The quotient in a division equation is like finding the unknown factor in a multiplication equation.

   The blanks in John's two equations represent the number of groups. Draw an array to represent the equations.

   The answer to both of John's equations is 7. I know 7 represents the number of groups, so I can draw 7 rows in my array. Then I can draw 5 dots in each row to show the size of the group for a total of 35 dots in my array.

   Lesson 6: Interpret the unknown in division using the array model.

Name _____ Date _____

1. Mr. Hannigan puts 12 pencils into boxes. Each box holds 4 pencils. Circle groups of 4 to show the pencils in each box.

Mr. Hannigan needs _____ boxes.

_____ × 4 = 12

12 ÷ 4 = _____

2. Mr. Hannigan places 12 pencils into 3 equal groups. Draw to show how many pencils are in each group.

There are _____ pencils in each group.

3 × _____ = 12

12 ÷ 3 = _____

3. Use an array to model Problem 1.

    a. _____ × 4 = 12

       12 ÷ 4 = _____

       The number in the blanks represents _____.

    b. 3 × _____ = 12

       12 ÷ 3 = _____

       The number in the blanks represents _____.

Lesson 6: Interpret the unknown in division using the array model.

4. Judy washes 24 dishes. She then dries and stacks the dishes equally into 4 piles. How many dishes are in each pile?

   24 ÷ 4 = _____

   4 × _____ = 24

   What is the meaning of the unknown factor and quotient? _____

5. Nate solves the equation _____ × 5 = 15 by writing and solving 15 ÷ 5 = ____. Explain why Nate's method works.

6. The blanks in Problem 5 represent the number of groups. Draw an array to represent the equations.

1. Draw an array that shows 5 rows of 2.

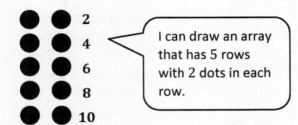

I can draw an array that has 5 rows with 2 dots in each row.

2. Draw an array that shows 2 rows of 5.

I can draw an array that has 2 rows with 5 dots in each row.

Write a multiplication sentence where the first factor represents the number of rows.

___5___ × ___2___ = ___10___

I can write a multiplication sentence with 5 as the first factor because 5 is the number of rows. The second factor is 2 because there are 2 dots in each row. I can skip-count by 2 to find the product, 10.

Write a multiplication sentence where the first factor represents the number of rows.

___2___ × ___5___ = ___10___

I can write a multiplication sentence with 2 as the first factor because 2 is the number of rows. The second factor is 5 because there are 5 dots in each row. I can skip-count by 5 to find the product, 10.

3. Why are the factors in your multiplication sentences in a different order?

The factors are in a different order because they mean different things. Problem 1 is 5 rows of 2, and Problem 2 is 2 rows of 5. In Problem 1, the 5 represents the number of rows. In Problem 2, the 5 represents the number of dots in each row.

The arrays show the commutative property. The order of the factors changed because the factors mean different things for each array. The product stayed the same for each array.

4. Write a multiplication sentence to match the number of groups. Skip-count to find the totals.

   a. 7 twos: $\quad 7 \times 2 = 14$

   b. 2 sevens: $\quad 2 \times 7 = 14$

   > 7 twos is unit form. It means that there are 7 groups of 2. I can represent that with the multiplication equation $7 \times 2 = 14$. 2 sevens means 2 groups of 7, which I can represent with the multiplication equation $2 \times 7 = 14$.

   > I see a pattern! 7 twos is equal to 2 sevens. It's the commutative property! The factors switched places and mean different things, but the product didn't change.

5. Find the unknown factor to make each equation true.

   $2 \times 8 = 8 \times \underline{2}$ $\qquad$ $\underline{4} \times 2 = 2 \times 4$

   > To make true equations, I need to make sure what's on the left of the equal sign is the same as (or equal to) what's on the right of the equal sign.

   > I can use the commutative property to help me. I know that $2 \times 8 = 16$ and $8 \times 2 = 16$, so I can write 2 in the first blank. To solve the second problem, I know that $4 \times 2 = 8$ and $2 \times 4 = 8$. I can write 4 in the blank.

Lesson 7: Demonstrate the commutativity of multiplication, and practice related facts by skip-counting objects in array models.

Name _____   Date _____

1. a. Draw an array that shows 7 rows of 2.

   b. Write a multiplication sentence where the first factor represents the number of rows.

   _____ × _____ = _____

2. a. Draw an array that shows 2 rows of 7.

   b. Write a multiplication sentence where the first factor represents the number of rows.

   _____ × _____ = _____

3. a. Turn your paper to look at the arrays in Problems 1 and 2 in different ways. What is the same and what is different about them?

   b. Why are the factors in your multiplication sentences in a different order?

4. Write a multiplication sentence to match the number of groups. Skip-count to find the totals. The first one is done for you.

   a. 2 twos: 2 × 2 = 4

   b. 3 twos: _____

   c. 2 threes: _____

   d. 2 fours: _____

   e. 4 twos: _____

   f. 5 twos: _____

   g. 2 fives: _____

   h. 6 twos: _____

   i. 2 sixes: _____

Lesson 7: Demonstrate the commutativity of multiplication, and practice related facts by skip-counting objects in array models.

5. Write and solve multiplication sentences where the second factor represents the size of the row.

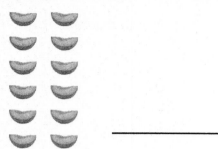   

_____    _____

6. Angel writes 2 × 8 = 8 × 2 in his notebook. Do you agree or disagree? Draw arrays to help explain your thinking.

7. Find the missing factor to make each equation true.

| 2 × 6 = 6 × _____ | _____ × 2 = 2 × 7 | 9 × 2 = _____ × 9 | 2 × _____ = 10 × 2 |

8. Tamia buys 2 bags of candy. Each bag has 7 pieces of candy in it.
   a. Draw an array to show how many pieces of candy Tamia has altogether.

   b. Write and solve a multiplication sentence to describe the array.

   c. Use the commutative property to write and solve a different multiplication sentence for the array.

A STORY OF UNITS – TEKS EDITION

Lesson 8 Homework Helper   3•1

1. Find the unknowns that make the equations true. Then, draw a line to match related facts.

   a. $3 + 3 + 3 + 3 =$ __12__

   b. $3 \times 7 =$ __21__

   c. $5 \text{ threes} + 1 \text{ three} =$ __6 threes__

   d. $3 \times 6 =$ __18__

   e. __12__ $= 4 \times 3$

   f. $21 = 7 \times$ __3__

   > $3 + 3 + 3 + 3$ is the same as 4 threes or 4×3, which equals 12. These equations are related because they both show that 4 groups of 3 equal 12.

   > 5 threes + 1 three = 6 threes. 6 threes is the same as 6 groups of 3 or 6×3, which equals 18. I can use the commutative property to match this equation with $3 \times 6 = 18$.

   > I can use the commutative property to match $3 \times 7 = 21$ and $21 = 7 \times 3$.

2. Fred puts 3 stickers on each page of his sticker album. He puts stickers on 7 pages.

   a. Use circles to draw an array that represents the total number of stickers in Fred's sticker album.

   ```
   ● ● ●   3
   ● ● ●   6
   ● ● ●   9
   ● ● ●   12
   ● ● ●   15
   ● ● ●   18
   ● ● ●   21
   ✗ ✗ ✗
   ✗ ✗ ✗
   ✗ ✗ ✗
   ```

   > I can draw an array with 7 rows to represent the 7 pages of the sticker album. I can draw 3 circles in each row to represent the 3 stickers that Fred puts on each page.

   > I can draw 3 more rows of 3 to represent the 3 pages and 3 stickers on each page that Fred adds to his sticker album in part (c).

Lesson 8: Demonstrate the commutativity of multiplication, and practice related facts by skip-counting objects in array models.

b. Use your array to write and solve a multiplication sentence to find Fred's total number of stickers.

7 × 3 = 21

Fred puts 21 stickers in his sticker album.

> I can write the multiplication equation 7 × 3 = 21 to find the total because there are 7 rows in my array with 3 circles in each row. I can use my array to skip-count to find the total, 21.

c. Fred adds 3 more pages to his sticker album. He puts 3 stickers on each new page. Draw x's to show the new stickers on the array in part (a).

d. Write and solve a multiplication sentence to find the new total number of stickers in Fred's sticker album.

24, 27, 30

10 × 3 = 30

Fred has a total of 30 stickers in his sticker album.

> I can continue to skip-count by three from 21 to find the total, 30. I can write the multiplication equation 10 × 3 = 30 to find the total because there are 10 rows in my array with 3 in each row. The number of rows changed, but the size of each row stayed the same.

Lesson 8: Demonstrate the commutativity of multiplication, and practice related facts by skip-counting objects in array models.

Name _____ Date _____

1. Draw an array that shows 6 rows of 3.

2. Draw an array that shows 3 rows of 6.

3. Write multiplication expressions for the arrays in Problems 1 and 2. Let the first factor in each expression represent the number of rows. Use the commutative property to make sure the equation below is true.

_____ × _____ = _____ × _____
      Problem 1   **Problem 2**

4. Write a multiplication sentence for each expression. You might skip-count to find the totals. The first one is done for you.

   a. 5 threes: __5 × 3 = 15__   d. 3 sixes: _____   g. 8 threes: _____

   b. 3 fives: _____   e. 7 threes: _____   h. 3 nines: _____

   c. 6 threes: _____   f. 3 sevens: _____   i. 10 threes: _____

5. Find the unknowns that make the equations true. Then, draw a line to match related facts.

   a. 3 + 3 + 3 + 3 + 3 + 3 = _____   d. 3 × 9 = _____

   b. 3 × 5 = _____   e. _____ = 6 × 3

   c. 8 threes + 1 three = _____   f. 15 = 5 × _____

Lesson 8: Demonstrate the commutativity of multiplication, and practice related facts by skip-counting objects in array models.

A STORY OF UNITS – TEKS EDITION
Lesson 8 Homework 3•1

6. Fernando puts 3 pictures on each page of his photo album. He puts pictures on 8 pages.

   a. Use circles to draw an array that represents the total number of pictures in Fernando's photo album.

   b. Use your array to write and solve a multiplication sentence to find Fernando's total number of pictures.

   c. Fernando adds 2 more pages to his book. He puts 3 pictures on each new page. Draw x's to show the new pictures on the array in Part (a).

   d. Write and solve a multiplication sentence to find the new total number of pictures in Fernando's album.

7. Ivania recycles. She gets 3 cents for every can she recycles.

   a. How much money does Ivania make if she recycles 4 cans?

   _____ × _____ = _____ cents

   b. How much money does Ivania make if she recycles 7 cans?

   _____ × _____ = _____ cents

Lesson 8: Demonstrate the commutativity of multiplication, and practice related facts by skip-counting objects in array models.

Lesson 9 Homework Helper 3•1

1. Matt organizes his baseball cards into 3 rows of three. Jenna adds 2 more rows of 3 baseball cards. Complete the equations to describe the total number of baseball cards in the array.

a. $(3 + 3 + 3) + (3 + 3) =$ __15__

b. 3 threes + __2__ threes = __5__ threes

c. __5__ $\times 3 =$ __15__

> The multiplication equation for this array is $5 \times 3 = 15$ because there are 5 threes or 5 rows of 3, which is a total of 15 baseball cards.

> The total for Matt's baseball cards (the unshaded rectangles) can be represented by $3 + 3 + 3$ because there are 3 rows of 3 baseball cards. The total for Jenna's baseball cards (the shaded rectangles) can be represented by $3 + 3$ because there are 2 rows of 3 baseball cards. This can be represented in unit form with 3 threes + 2 threes, which equals 5 threes.

2. $8 \times 3 =$ __24__

> I can find the product of 8×3 using the array and the equations below. This problem is different than the problem above because now I am finding two products and subtracting instead of adding.

$10 \times 3 =$ __30__

$2 \times 3 =$ __6__

> The multiplication equation for the whole array is $10 \times 3 = 30$. The multiplication equation for the shaded part is $2 \times 3 = 6$.

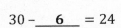
30 − __6__ = 24

__8__ $\times 3 = 24$

> To solve 8×3, I can think of 10×3 because that's an easier fact. I can subtract the product of 2×3 from the product of 10×3. $30 − 6 = 24$, so $8 \times 3 = 24$.

Lesson 9: Find related multiplication facts by adding and subtracting equal groups in array models.

Name _____ Date _____

1. Dan organizes his stickers into 3 rows of four. Irene adds 2 more rows of stickers. Complete the equations to describe the total number of stickers in the array.

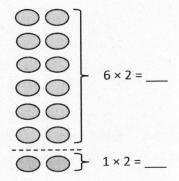

a. (4 + 4 + 4) + (4 + 4) = _____

b. 3 fours + _____ fours = _____ fours

c. _____ × 4 = _____

2. 7 × 2 = _____

6 × 2 = ___

1 × 2 = ___

12 + 2 = _____

_____ × 2 = 14

3. 9 × 3 = _____

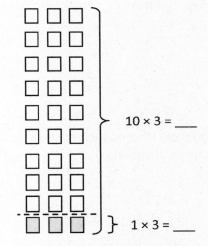

10 × 3 = ___

1 × 3 = ___

30 − _____ = 27

_____ × 3 = 27

Lesson 9: Find related multiplication facts by adding and subtracting equal groups in array models.

4. Franklin collects stickers. He organizes his stickers in 5 rows of four.

   a. Draw an array to represent Franklin's stickers. Use an x to show each sticker.

   b. Solve the equation to find Franklin's total number of stickers. 5 × 4 = _____

5. Franklin adds 2 more rows. Use circles to show his new stickers on the array in Problem 4(a).

   a. Write and solve an equation to represent the circles you added to the array.

   _____ × 4 = _____

   b. Complete the equation to show how you add the totals of 2 multiplication facts to find Franklin's total number of stickers.

   _____ + _____ = 28

   c. Complete the unknown to show Franklin's total number of stickers.

   _____ × 4 = 28

Lesson 10 Homework Helper 3•1

1. Use the array to help you fill in the blanks.

   6 × 2 = __12__

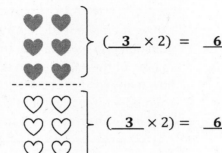

(__3__ × 2) = __6__

(__3__ × 2) = __6__

The dotted line in the array shows how I can break apart 6 × 2 into two smaller facts. Then I can add the products of the smaller facts to find the product of 6 × 2.

I know the first factor in each equation is 3 because there are 3 rows in each of the smaller arrays. The product for each array is 6.

(3 × 2) + (3 × 2) = __6__ + __6__

__6__ × 2 = __12__

The expressions in the parentheses represent the smaller arrays. I can add the products of these expressions to find the total number of hearts in the array. The products of the smaller expressions are both 6. 6 + 6 = 12, so 6 × 2 = 12.

Hey, look! It's a doubles fact! 6 + 6 = 12. I know my doubles facts, so this is easy to solve!

Lesson 10: Model the distributive property with arrays to decompose units as a strategy to multiply.

2. Lilly puts stickers on a piece of paper. She puts 3 stickers in each row.

   a. Fill in the equations to the right. Use them to draw arrays that show the stickers on the top and bottom parts of Lilly's paper.

I know there are 3 stickers in each row, and this equation also tells me that there are 12 stickers in all on the top of the paper. I can skip-count by 3 to figure out how many rows of stickers there. 3, 6, 9, 12. I skip-counted 4 threes, so there are 4 rows of 3 stickers. Now I can draw an array with 4 rows of 3.

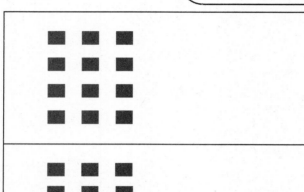

__4__ × 3 = 12

__2__ × 3 = 6

I see 6 rows of 3 altogether. I can use the products of these two smaller arrays to solve 6 × 3.

I can use the same strategy to find the number of rows in this equation. I skip-counted 2 threes, so there are 2 rows of 3 stickers. Now I can draw an array with 2 rows of 3.

Lesson 10: Model the distributive property with arrays to decompose units as a strategy to multiply.

Name _____ Date _____

1. 6 × 3 = _____

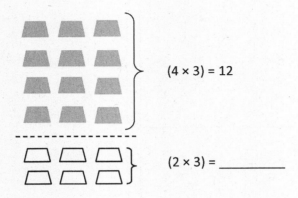

(4 × 3) = 12

(2 × 3) = _____

12 + _____ = _____

6 × 3 = _____

2. 8 × 2 = _____

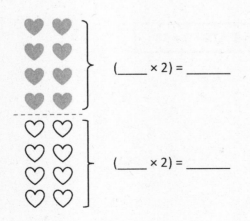

(___ × 2) = _____

(___ × 2) = _____

(4 × 2) + (4 × 2) = _____ + _____

___ × 2 = _____

Lesson 10: Model the distributive property with arrays to decompose units as a strategy to multiply.

3. Adriana organizes her books on shelves. She puts 3 books in each row.

   a. Fill in the equations on the right. Use them to draw arrays that show the books on Adriana's top and bottom shelves.

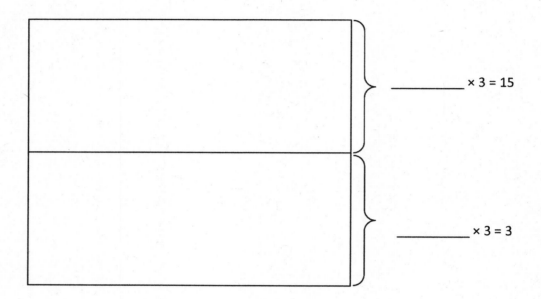

_____ × 3 = 15

_____ × 3 = 3

   b. Adriana calculates the total number of books as shown below. Use the array you drew to help explain Adriana's calculation.

   $$6 \times 3 = 15 + 3 = 18$$

1. Mr. Russell organizes 18 clipboards equally into 3 boxes. How many clipboards are in each box? Model the problem with both an array and a labeled strip diagram. Show each column as the number of clipboards in each box.

I can draw an array with 3 columns because each column represents 1 box of clipboards. I can draw rows of 3 dots until I have a total of 18 dots. I can count how many dots are in each column to solve the problem.

I know the total number of clipboards is 18, and there are 3 boxes of clipboards. I need to figure out how many clipboards are in each box. I can think of this as division, $18 \div 3 = \underline{}$, or as multiplication, $3 \times \underline{} = 18$.

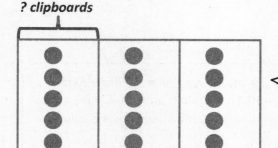

? clipboards

18 *clipboards*

I can draw 3 units in my strip diagram to represent the 3 boxes of clipboards. I can label the whole strip diagram with "18 clipboards". I can label one unit in the strip diagram with "? clipboards" because that's what I am solving for. I can draw 1 dot in each unit until I have a total of 18 dots.

There are __6__ clipboards in each box.

Look, my array and strip diagram both show units of 6. The columns in my array each have 6 dots, and the units in my strip diagram each have a value of 6.

I know the answer is 6 because my array has 6 dots in each column. My strip diagram also shows the answer because there are 6 dots in each unit.

Lesson 11: Model division as the unknown factor in multiplication using arrays and strip diagrams.

2. Caden reads 2 pages in his book each day. How many days will it take him to read a total of 12 pages?

This problem is different than the other problem because the known information is the total and the size of each group. I need to figure out how many groups there are.

I can draw an array where each column represents the number of pages Caden reads each day. I can keep drawing columns of 2 until I have a total of 12.

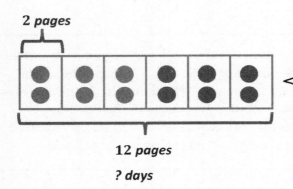

I can use my array to help me draw a strip diagram. I can draw 6 units of 2 in my strip diagram because my array shows 6 columns of 2.

$12 \div 2 = 6$

I know the answer is 6 because my array shows 6 columns of 2, and my strip diagram shows 6 units of 2.

It will take Caden 6 days to read a total of 12 pages.

I can write a statement to answer the question.

Name _____ Date _____

1. Fred has 10 pears. He puts 2 pears in each basket. How many baskets does he have?

    a. Draw an array where each column represents the number of pears in each basket.

    _____ ÷ 2 = _____

    b. Redraw the pears in each basket as a unit in the strip diagram. Label the diagram with known and unknown information from the problem.

    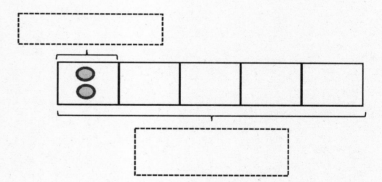

2. Ms. Meyer organizes 15 clipboards equally into 3 boxes. How many clipboards are in each box? Model the problem with both an array and a labeled strip diagram. Show each column as the number of clipboards in each box.

    There are _____ clipboards in each box.

3. Sixteen action figures are arranged equally on 2 shelves. How many action figures are on each shelf? Model the problem with both an array and a labeled strip diagram. Show each column as the number of action figures on each shelf.

4. Jasmine puts 18 hats away. She puts an equal number of hats on 3 shelves. How many hats are on each shelf? Model the problem with both an array and a labeled strip diagram. Show each column as the number of hats on each shelf.

5. Corey checks out 2 books a week from the library. How many weeks will it take him to check out a total of 14 books?

Lesson 12 Homework Helper   3•1

1. Mrs. Harris divides 14 flowers equally into 7 groups for students to study. Draw flowers to find the number in each group. Label known and unknown information on the strip diagram to help you solve.

> I know the total number of flowers and the number of groups. I need to solve for the number of flowers in each group.

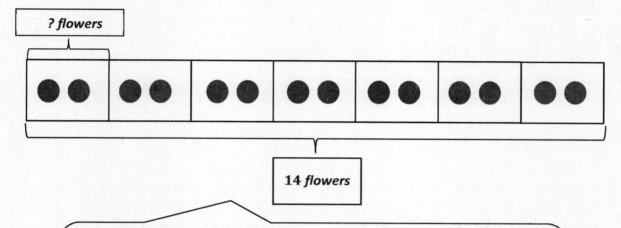

> I can label the value of the strip diagram as "14 flowers". The number of units in the strip diagram, 7, represents the number of groups. I can label the unknown, which is the value of each unit, as "? flowers". I can draw 1 flower in each unit until I have a total of 14 flowers. I can draw dots instead of flowers to be more efficient!

> I can use my strip diagram to solve the problem by counting the number of dots in each unit.

$7 \times \underline{\ \ 2\ \ } = 14$

$14 \div 7 = \underline{\ \ 2\ \ }$

There are __2__ flowers in each group.

Lesson 12: Interpret the quotient as the number of groups or the number of objects in each group using units of 2.

2. Lauren finds 2 rocks each day for her rock collection. How many days will it take Lauren to find 16 rocks for her rock collection?

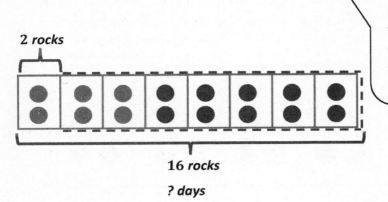

I know the total is 16 rocks. I know Lauren finds 2 rocks each day, which is the size of each group. I need to figure out how many days it will take her to collect 16 rocks. The unknown is the number of groups.

I can draw a strip diagram to solve this problem. I can draw a unit of 2 to represent the 2 rocks that Lauren collects each day. I can draw a dotted line to estimate the total days. I can draw units of 2 until I have a total of 16 rocks. I can count the number of units to find the answer.

$16 \div 2 = 8$

I know the answer is 8 because my strip diagram shows 8 units of 2.

It will take Lauren 8 days to find 16 rocks.

I can write a statement to answer the question.

Name _____ Date _____

1. Ten people wait in line for the roller coaster. Two people sit in each car. Circle to find the total number of cars needed.

10 ÷ 2 = _____

There are _____ cars needed.

2. Mr. Ramirez divides 12 frogs equally into 6 groups for students to study. Draw frogs to find the number in each group. Label known and unknown information on the strip diagram to help you solve.

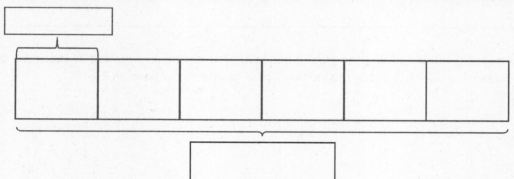

6 × _____ = 12

12 ÷ 6 = _____

There are _____ frogs in each group.

3. Match.

4. Betsy pours 16 cups of water to equally fill 2 bottles. How many cups of water are in each bottle? Label the strip diagram to represent the problem, including the unknown.

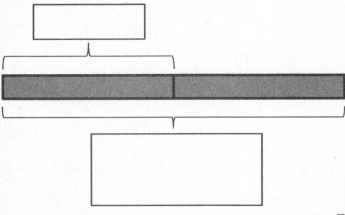

There are _____ cups of water in each bottle.

5. An earthworm tunnels 2 centimeters into the ground each day. The earthworm tunnels at about the same pace every day. How many days will it take the earthworm to tunnel 14 centimeters?

6. Seb and Shawn go to the movies. The tickets cost $16 in total. The boys share the cost equally. How much does Shawn pay?

A STORY OF UNITS – TEKS EDITION   Lesson 13 Homework Helper 3•1

1. Mr. Stroup's pet fish are shown below. He keeps 3 fish in each tank.
   a. Circle to show how many fish tanks he has. Then, skip-count to find the total number of fish.

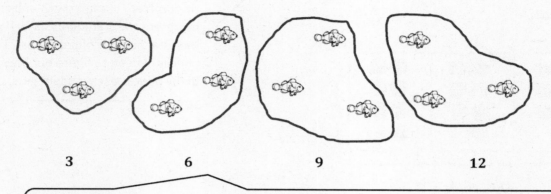

   3   6   9   12

   > I can circle groups of 3 fish and skip-count by 3 to find the total number of fish. I can count the number of groups to figure out how many fish tanks Mr. Stroup has.

   Mr. Stroup has a total of 12 fish in 4 tanks.

   b. Draw and label a strip diagram to represent the problem.

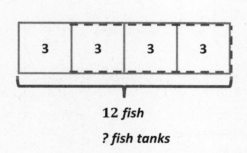

   12 *fish*
   ? *fish tanks*

   > I can use the picture in part (a) to help me draw a strip diagram. Each fish tank has 3 fish, so I can label each unit with the number 3. I can draw a dotted line to estimate the total fish tanks. I can label the total as 12 fish. Then I can draw units of 3 until I have a total of 12 fish.

   > The picture and the strip diagram both show that there are 4 fish tanks. The picture shows 4 equal groups of 3, and the strip diagram shows 4 units of 3.

   __12__ ÷ 3 = __4__

   Mr. Stroup has __4__ fish tanks.

Lesson 13: Interpret the quotient as the number of groups or the number of objects in each group using units of 3.

2. A teacher has 21 pencils. They are divided equally among 3 students. How many pencils does each student get?

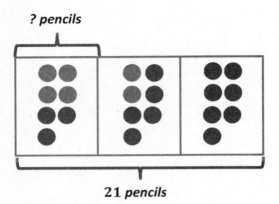

> I can draw a strip diagram to solve this problem. I can draw 3 units to represent the 3 students. I can label the total number of pencils as 21 pencils. I need to figure out how many pencils each student gets.

> I know that I can divide 21 by 3 to solve. I don't know 21 ÷ 3, so I can draw one dot in each unit until I have a total of 21 dots. I can count the number of dots in one unit to find the quotient.

21 ÷ 3 = 7

> I know the answer is 7 because my strip diagram shows 3 units of 7.

Each student will get 7 pencils.

> I can write a statement to answer the question.

Name _____ Date _____

1. Fill in the blanks to make true number sentences.

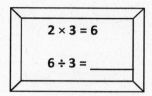
2 × 3 = 6
6 ÷ 3 = _____

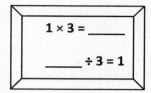
1 × 3 = _____
_____ ÷ 3 = 1

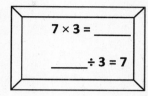
7 × 3 = _____
_____ ÷ 3 = 7

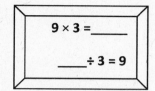
9 × 3 = _____
_____ ÷ 3 = 9

2. Ms. Gillette's pet fish are shown below. She keeps 3 fish in each tank.

   a. Circle to show how many fish tanks she has. Then, skip-count to find the total number of fish.

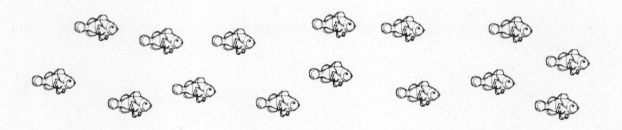

   b. Draw and label a strip diagram to represent the problem.

   _____ ÷ 3 = _____

   Ms. Gillette has _____ fish tanks.

3. Juan buys 18 meters of wire. He cuts the wire into pieces that are each 3 meters long. How many pieces of wire does he cut?

4. A teacher has 24 pencils. They are divided equally among 3 students. How many pencils does each student get?

5. There are 27 third-graders working in groups of 3. How many groups of third-graders are there?

1. Mrs. Smith replaces 4 wheels on 3 cars. How many wheels does she replace? Draw and label a strip diagram to solve.

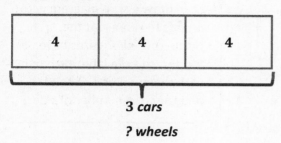

3 cars

? wheels

4, 8, 12

3 × 4 = 12

> I can draw a strip diagram with 3 units to represent the 3 cars. Each car has 4 wheels, so I can label each unit with the number 4. I need to find the total number of wheels.

> I can skip-count by fours or multiply 3 × 4 to find how many wheels Mrs. Smith replaces.

Mrs. Smith replaces ___12___ wheels

2. Thomas makes 4 necklaces. Each necklace has 7 beads. Draw and label a strip diagram to show the total number of beads Thomas uses.

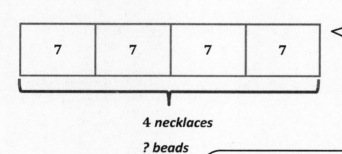

4 necklaces

? beads

7, 14, 21, 28

4, 8, 12, 16, 20, 24, 28

4 × 7 = 28

> I can draw a strip diagram with 4 units to represent the 4 necklaces. I can label each unit in the strip diagram to show that every necklace has 7 beads. I need to find the total number of beads.

> I can skip-count 4 sevens, but sevens are still tricky for me. I can skip-count 7 fours instead! I can also multiply 4 × 7 to find how many beads Thomas uses.

Thomas uses ___28___ beads

Lesson 14: Skip-count objects in models to build fluency with multiplication facts using units of 4.

3. Find the total number of sides on 6 squares.

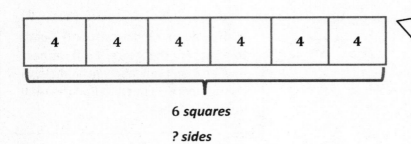

6 *squares*

? *sides*

> I can draw a strip diagram with 6 units to represent the 6 squares. All squares have 4 sides, so I can label each unit with the number 4. I need to find the total number of sides.

4, 8, 12, 16, 20, 24

> I can skip-count 6 fours or multiply 6×4 to find the total number of sides on 6 squares.

$6 \times 4 = 24$

There are 24 sides on 6 squares.

Name _____ Date _____

1. Skip-count by fours. Match each answer to the appropriate expression.

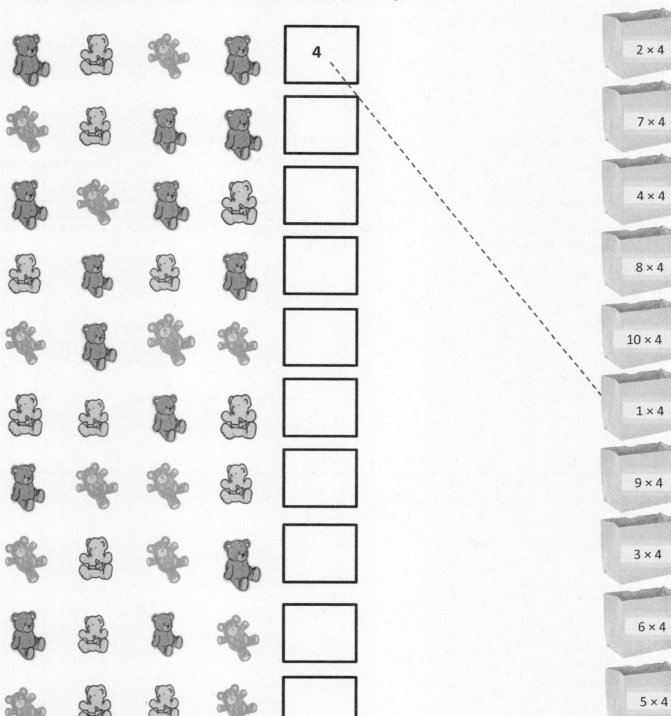

Lesson 14: Skip-count objects in models to build fluency with multiplication facts using units of 4.

2. Lisa places 5 rows of 4 juice boxes in the refrigerator. Draw an array and skip-count to find the total number of juice boxes.

There are _____ juice boxes in total.

3. Six folders are placed on each table. How many folders are there on 4 tables? Draw and label a strip diagram to solve.

4. Find the total number of corners on 8 squares.

1. Label the strip diagrams, and complete the equations. Then, draw an array to represent the problems.

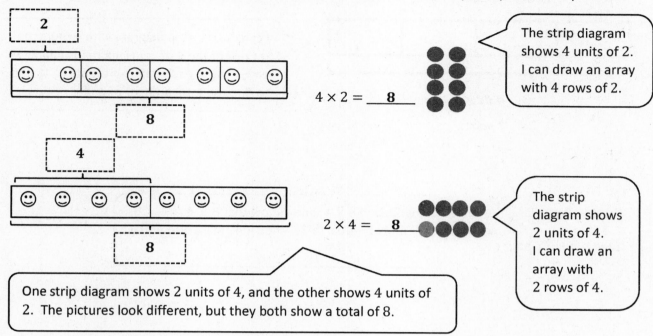

2. 8 books cost $4 each. Draw and label a strip diagram to show the total cost of the books.

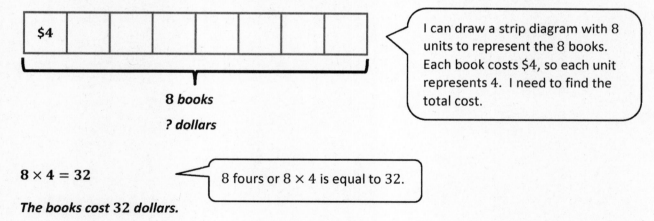

$8 \times 4 = 32$

8 fours or 8×4 is equal to 32.

The books cost 32 dollars.

3. Liana reads 8 pages from her book each day. How many pages does Liana read in 4 days?

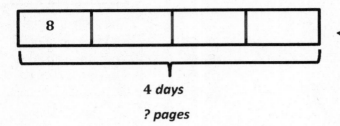

4 days

? pages

I can draw a strip diagram with 4 units to represent the 4 days. Liana reads 8 pages each day, so each unit represents 8. I need to find the total number of pages.

$4 \times 8 = 32$

Liana reads 32 pages.

I just solved 8×4, and I know that $8 \times 4 = 4 \times 8$. If 8 fours is equal to 32, then 4 eights is also equal to 32.

Name _____ Date _____

1. Label the strip diagrams and complete the equations. Then, draw an array to represent the problems.

a.

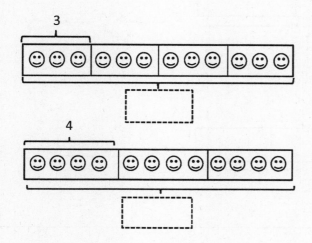

4 × 3 = _____

3 × 4 = _____

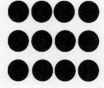

b.

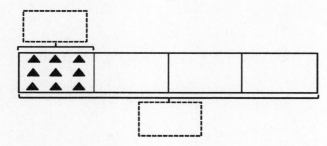

4 × _____ = _____

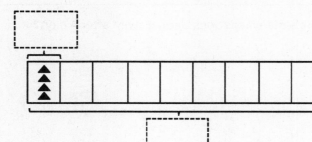

_____ × 4 = _____

Lesson 15: Relate arrays to strip diagrams to model the commutative property of multiplication.

c.

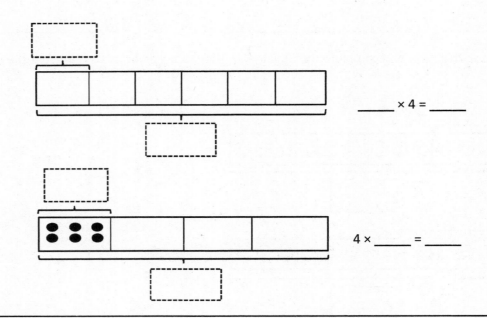

____ × 4 = ____

4 × ____ = ____

2. Seven clowns hold 4 balloons each at the fair. Draw and label a strip diagram to show the total number of balloons the clowns hold.

3. George swims 7 laps in the pool each day. How many laps does George swim after 4 days?

1. Label the array. Then, fill in the blanks below to make true number sentences.

$8 \times 3 = \underline{24}$

$(5 \times 3) = \underline{15}$

$(3 \times 3) = \underline{9}$

> I know that I can break apart 8 threes into 5 threes and 3 threes. I can add the products for 5×3 and 3×3 to find the product for 8×3.

$$8 \times 3 = (5 \times 3) + (3 \times 3)$$
$$= \underline{15} + \underline{9}$$
$$= \underline{24}$$

2. The array below shows one strategy for solving 8×4. Explain the strategy using your own words.

 $(5 \times 4) = \underline{20}$

 $(3 \times 4) = \underline{12}$

> 8×4 is a tricky fact for me to solve, but 5×4 and 3×4 are both pretty easy facts. I can use them to help me!

I split apart the 8 rows of 4 into 5 rows of 4 and 3 rows of 4. I split the array there because my fives facts and my threes facts are easier than my eights facts. I know that $5 \times 4 = 20$ and $3 \times 4 = 12$. I can add those products to find that $8 \times 4 = 32$.

Lesson 16: Use the distributive property as a strategy to find related multiplication facts.

Name _____ Date _____

1. Label the array. Then, fill in the blanks below to make true number sentences.

   a. **6 × 4** = _____

   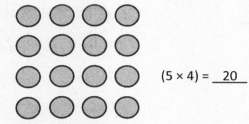

   (5 × 4) = __20__

   (___ × 4) = _____

   (6 × 4) = (5 × 4) + (___ × 4)

            = __20__ + _____

            = _____

   b. **8 × 4** = _____

   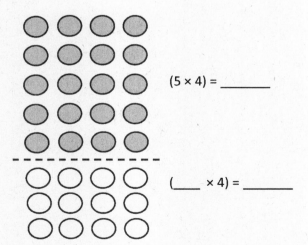

   (5 × 4) = _____

   (___ × 4) = _____

   (8 × 4) = (5 × 4) + (___ × 4)

            = _____ + _____

            = _____

Lesson 16: Use the distributive property as a strategy to find related multiplication facts.

2. Match the multiplication expressions with their answers.

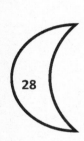 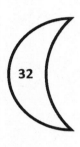  

3. The array below shows one strategy for solving 9 × 4. Explain the strategy using your own words.

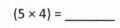

(5 × 4) = _____

(4 × 4) = _____

Lesson 16: Use the distributive property as a strategy to find related multiplication facts.

1. The baker packs 20 muffins into boxes of 4. Draw and label a strip diagram to find the number of boxes she packs.

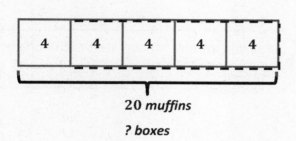

20 *muffins*

? *boxes*

$20 \div 4 = \underline{5}$

The baker packs 5 boxes.

> I can draw a strip diagram. Each box has 4 muffins, so I can draw a unit and label it 4. I can draw a dotted line to estimate the total number of boxes, because I don't yet know how many boxes there are. I do know the total, so I'll label that as 20 muffins. I'll solve by drawing units of 4 in the dotted part of my strip diagram until I have a total of 20 muffins. Then I can count the number of units to see how many boxes of muffins the baker packs.

2. The waiter arranges 12 plates into 4 equal rows. How many plates are in each row?

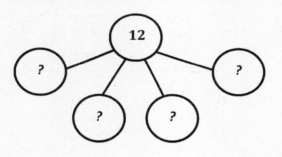

$12 \div 4 = \underline{3}$

$3 \times 4 = \underline{12}$

There are 3 plates in each row.

> I can use a number bond to solve. I know that the total number of plates is 12 and that the 12 plates are in 4 rows. Each part in the number bond represents a row of plates.

> I can divide to solve. I can also think of this as multiplication with an unknown factor.

Lesson 17: Model the relationship between multiplication and division.

3. A teacher has 20 erasers. She divides them equally between 4 students. She finds 12 more erasers and divides these equally between the 4 students as well. How many erasers does each student receive?

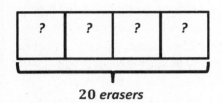

$20 \div 4 = \underline{\ 5\ }$

> I can find the number of erasers each student gets at first when the teacher has 20 erasers.

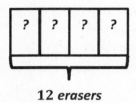

$12 \div 4 = \underline{\ 3\ }$

> I can find how many erasers each student gets when the teacher finds 12 more erasers.

5 erasers + 3 erasers = $\underline{\ 8\ }$ erasers.

Each student receives 8 erasers.

> I can add to find how many total erasers each student gets.

Name _____ Date _____

1. Use the array to complete the related equations.

   $1 \times 4 = $ _____   _____ $\div 4 = 1$

   $2 \times 4 = $ _____   _____ $\div 4 = 2$

   _____ $\times 4 = 12$ $12 \div 4 = $ _____

   _____ $\times 4 = 16$ $16 \div 4 = $ _____

   _____ $\times$ _____ $= 20$ $20 \div$ _____ $=$ _____

   _____ $\times$ _____ $= 24$ $24 \div$ _____ $=$ _____

   _____ $\times 4 = $ _____   _____ $\div 4 = $ _____

   _____ $\times 4 = $ _____   _____ $\div 4 = $ _____

   _____ $\times$ _____ $=$ _____   _____ $\div$ _____ $=$ _____

   _____ $\times$ _____ $=$ _____   _____ $\div$ _____ $=$ _____

Lesson 17: Model the relationship between multiplication and division.

2. The teacher puts 32 students into groups of 4. How many groups does she make? Draw and label a strip diagram to solve.

3. The store clerk arranges 24 toothbrushes into 4 equal rows. How many toothbrushes are in each row?

4. An art teacher has 40 paintbrushes. She divides them equally among her 4 students. She finds 8 more brushes and divides these equally among the students, as well. How many brushes does each student receive?

1. Match the number bond on an apple with the equation on a bucket that shows the same total.

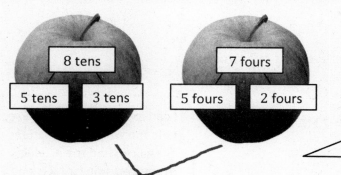

The number bonds in the apples help me see how I can find the total by adding the two smaller parts together. I can match the apples with the equations below that show the same two parts and total.

$(5 \times 4) + (2 \times 4) = 28$

$(5 \times 10) + (3 \times 10) = 80$

2. Solve.

$9 \times 4 = \underline{\ 36\ }$

I can think of this total as 9 fours. There are many ways to break apart 9 fours, but I'm going to break it apart as 5 fours and 4 fours because 5 is a friendly number.

I can use the number bond to help me fill in the blanks. Adding the **products** of these two smaller facts helps me find the product of the larger fact.

$(\underline{\ 5\ } \times 4) + (\underline{\ 4\ } \times 4) = 9 \times 4$

$\underline{\ 20\ } + \underline{\ 16\ } = \underline{\ 36\ }$

$9 \times 4 = \underline{\ 36\ }$

Lesson 18: Apply the distributive property to decompose units.

3. Mia solves 7 × 3 using the break apart and distribute strategy. Show an example of what Mia's work might look like below.

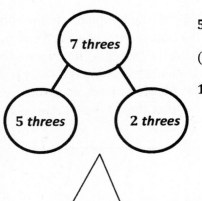

5 *threes* + 2 *threes* = 7 *threes*

(5 × 3) + (2 × 3) = 7 × 3

15 + 6 = 21

> I can use the number bond to help me write the equations. Then I can find the products of the two smaller facts and add them to find the product of the larger fact.

> The number bond helps me see the break apart and distribute strategy easily. I can think of 7 × 3 as 7 threes. Then I can break it apart as 5 threes and 2 threes.

Lesson 18: Apply the distributive property to decompose units.

Name _____ Date _____

1. Match.

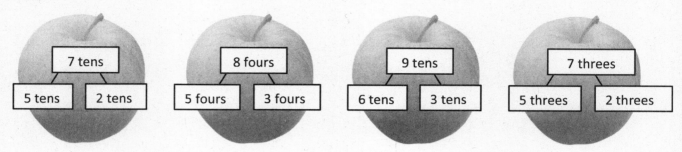

(5 × 4) + (3 × 4) = 32

(5 × 3) + (2 × 3) = 21

(5 × 10) + (2 × 10) = 70

(6 × 10) + (3 × 10) = 90

2. 9 × 4 = _____

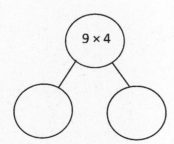

(_____ × 4) + (_____ × 4) = 9 × 4

_____ + _____ = _____

9 × 4 = _____

Lesson 18: Apply the distributive property to decompose units.

3. Lydia makes 10 pancakes. She tops each pancake with 4 blueberries. How many blueberries does Lydia use in all? Use the break apart and distribute strategy, and draw a number bond to solve.

Lydia uses _____ blueberries in all.

4. Steven solves 7 × 3 using the break apart and distribute strategy. Show an example of what Steven's work might look like below.

5. There are 7 days in 1 week. How many days are there in 10 weeks?

Lesson 18: Apply the distributive property to decompose units.

Lesson 19 Homework Helper 3•1

1. Solve.

    $28 \div 4 = \underline{7}$

    $(20 \div 4) = \underline{5}$

    $(8 \div 4) = \underline{2}$

    $(28 \div 4) = (20 \div 4) + (\underline{8} \div 4)$
    $= \underline{5} + \underline{2}$
    $= \underline{7}$

    This shows how we can add the quotients of two smaller facts to find the quotient of the larger one. The array can help me fill in the blanks.

    This array shows a total of 28 triangles. I see that the dotted line breaks apart the array after the fifth row. There are 5 fours above the dotted line and 2 fours below the dotted line.

Match equal expressions.

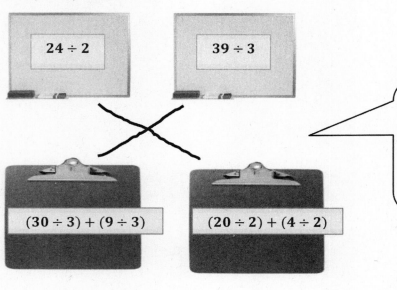

I can match the larger division problem found on the whiteboard to the two smaller division problems added together on the clipboard below.

Lesson 19: Apply the distributive property to decompose units.

2. Chloe draws the array below to find the answer to 48 ÷ 4. Explain Chloe's strategy.

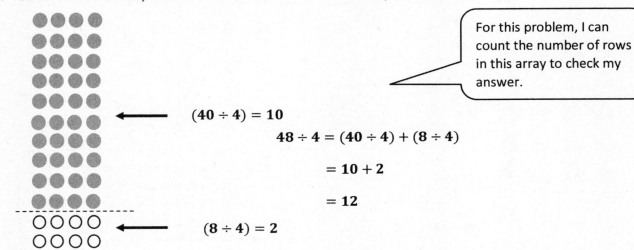

For this problem, I can count the number of rows in this array to check my answer.

$(40 \div 4) = 10$

$48 \div 4 = (40 \div 4) + (8 \div 4)$

$= 10 + 2$

$= 12$

$(8 \div 4) = 2$

Chloe breaks apart 48 as 10 fours and 2 fours. 10 fours equals 40, and 2 fours equals 8. So, she does 40 ÷ 4 and 8 ÷ 4 and adds the answers to get 48 ÷ 4, which equals 12.

Name _____ Date _____

1. Label the array. Then, fill in the blanks to make true number sentences.

a. 18 ÷ 3 = _____

(9 ÷ 3) = 3

(9 ÷ 3) = _____

(18 ÷ 3) = (9 ÷ 3) + (9 ÷ 3)
= __3__ + _____
= __6__

b. 21 ÷ 3 = _____

(15 ÷ 3) = 5

(6 ÷ 3) = _____

(21 ÷ 3) = (15 ÷ 3) + (6 ÷ 3)
= __5__ + _____
= _____

c. 24 ÷ 4 = _____

(20 ÷ 4) = _____

(4 ÷ 4) = _____

(24 ÷ 4) = (20 ÷ 4) + (____ ÷ 4)
= _____ + _____
= _____

d. 36 ÷ 4 = _____

(20 ÷ 4) = _____

(16 ÷ 4) = _____

(36 ÷ 4) = (____ ÷ 4) + (____ ÷ 4)
= _____ + _____
= _____

Lesson 19: Apply the distributive property to decompose units.

2. Match equal expressions.

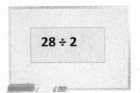   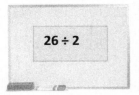

  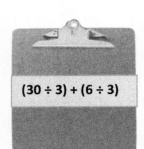 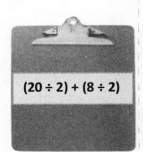

3. Alex draws the array below to find the answer to 35 ÷ 5. Explain Alex's strategy.

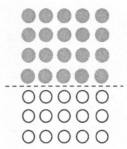

Lesson 20 Homework Helper

1. Thirty-five students are eating lunch at 5 tables. Each table has the same number of students.

   a. How many students are sitting at each table?

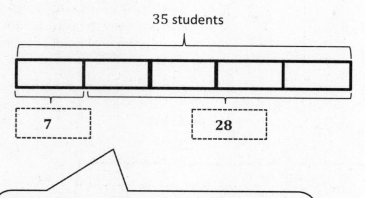

   I know there are a total of 35 students eating lunch at 5 tables. I know each table has the same number of students. I need to figure out how many students are sitting at each table. The unknown is the size of each group.

   Each unit in my strip diagram represents 1 table. Since there are 35 students and 5 tables, I can divide 35 by 5 to find that each table has 7 students. This strip diagram shows that there are 5 units of 7 for a total of 35.

   $35 \div 5 = 7$

   There are 7 students sitting at each table.

   b. How many students are sitting at 4 tables?

   $4 \times 7 = 28$
   There are 28 students sitting at 4 tables.

   I can write a number sentence and a statement to answer the question.

   Since I now know there are 7 students sitting at each table, I can multiply the number of tables, 4, by 7 to find that there are 28 students sitting at 4 tables. I can see this in the strip diagram: 4 units of 7 equal 28.

Lesson 20: Solve two-step word problems involving multiplication and division, and assess the reasonableness of answers.

2. The store has 30 notebooks in packs of 3. Six packs of notebooks are sold. How many packs of notebooks are left?

> I can draw a strip diagram that shows 30 notebooks in packs of 3. I can find the total number of packs by dividing 30 by 3 to get 10 total packs of notebooks.

> I know the total is 30 notebooks. I know the notebooks are in packs of 3. First I need to figure out how many total packs of notebooks are in the store.

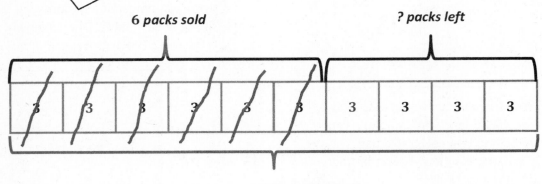

30 *notebooks*

? *total packs*

> Now that I know the total number of packs is 10, I can find the number of packs that are left.

$30 \div 3 = 10$

There are a total of 10 packs of notebooks at the store.

$10 - 6 = 4$

There are 4 packs of notebooks left.

> I can show the packs that were sold on my strip diagram by crossing off 6 units of 3. Four units of 3 are not crossed off, so there are 4 packs of notebooks left. I can write a subtraction equation to represent the work on my strip diagram.

Name _____ Date _____

1. Jerry buys a pack of pencils that costs $3. David buys 4 sets of markers. Each set of markers also costs $3.

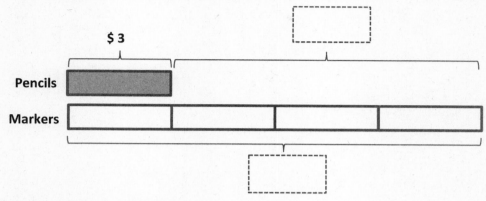

   a. What is the total cost of the markers?

   b. How much more does David spend on 4 sets of markers than Jerry spends on a pack of pencils?

2. Thirty students are eating lunch at 5 tables. Each table has the same number of students.

   a. How many students are sitting at each table?

   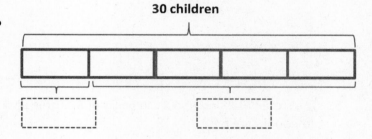

   b. How many students are sitting at 4 tables?

3. The teacher has 12 green stickers and 15 purple stickers. Three students are given an equal number of each color sticker. How many green and purple stickers does each student get?

4. Three friends go apple picking. They pick 13 apples on Saturday and 14 apples on Sunday. They share the apples equally. How many apples does each person get?

5. The store has 28 notebooks in packs of 4. Three packs of notebooks are sold. How many packs of notebooks are left?

Lesson 21 Homework Helper

1. John has a reading goal. He checks out 3 boxes of 7 books from the library. After finishing them, he realizes that he beat his goal by 5 books! Label the strip diagrams to find John's reading goal.

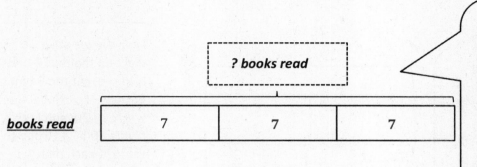

Each unit in this strip diagram represents 1 box of John's library books. The number of books in each box (the size) is 7 books. So I can multiply 3 × 7 to find the number of books John reads.

3 × 7 = 21
John reads 21 books.

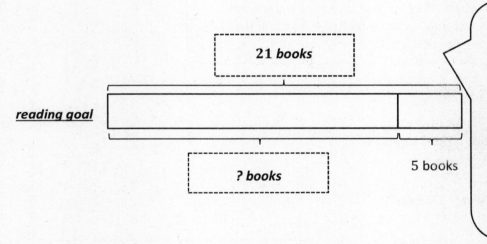

I can draw a strip diagram that shows 21 as the total because John reads 21 books. I can label one part as 5 because John beat his reading goal by 5 books. When I know a total and one part, I know I can subtract to find the other part.

21 − 5 = 16
John's goal was to read ___16___ books.

I can check back to see if my statement answers the question.

Lesson 21: Solve two-step word problems involving all four operations, and assess the reasonableness of answers.

2. Mr. Kim plants 20 trees around the neighborhood pond. He plants equal numbers of maple, pine, spruce, and birch trees. He waters the spruce and birch trees before it gets dark. How many trees does Mr. Kim still need to water? Draw and label a strip diagram.

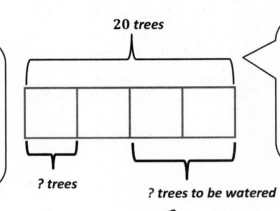

I know Mr. Kim plants a total of 20 trees. He plants an equal number of 4 types of trees. This is the number of groups. So, the unknown is the size of each group.

I can draw a strip diagram that has 4 units to represent the 4 types of trees. I can label the whole as 20, and I can divide 20 by 4 to find the value of each unit.

I know that Mr. Kim waters the spruce and birch trees, so he still needs to water the maple and pine trees. I can see from my strip diagram that 2 units of 5 trees still need to be watered. I can multiply 2 × 5 to find that 10 trees still need to be watered.

20 ÷ 4 = 5
Mr. Kim plants 5 of each type of tree.

2 × 5 = 10
Mr. Kim still needs to water 10 trees.

20 − 10 = 10
Mr. Kim still needs to water 10 trees.

Or I can subtract the number of trees watered, 10, from the total number of trees to find the answer.

Name _____ Date _____

1. Tina eats 8 crackers for a snack each day at school. On Friday, she drops 3 and only eats 5. Write and solve an equation to show the total number of crackers Tina eats during the week.

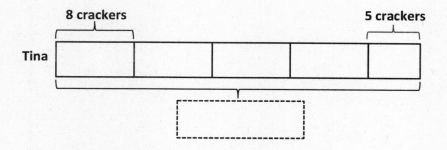

Tina eats _____ crackers.

2. Ballio has a reading goal. He checks 3 boxes of 9 books out from the library. After finishing them, he realizes that he beat his goal by 4 books! Label the strip diagrams to find Ballio's reading goal.

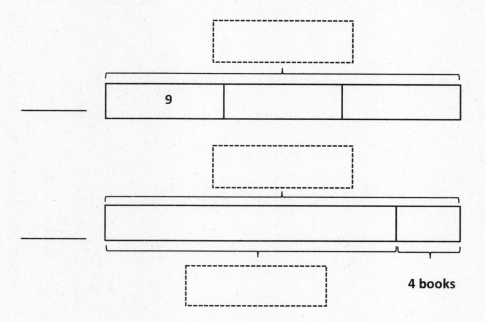

Ballio's goal is to read _____ books.

Lesson 21: Solve two-step word problems involving all four operations, and assess the reasonableness of answers.

3. Mr. Nguyen plants 24 trees around the neighborhood pond. He plants equal numbers of maple, pine, spruce, and birch trees. He waters the spruce and birch trees before it gets dark. How many trees does Mr. Nguyen still need to water? Draw and label a strip diagram.

4. Anna buys 24 seeds and plants 3 in each pot. She has 5 pots. How many more pots does Anna need to plant all of her seeds?

Grade 3
Module 2

The table to the right shows how much time it takes each of the 5 students to run 100 meters.

Eric	19 seconds
Woo	20 seconds
Sharon	24 seconds
Steven	18 seconds
Joyce	22 seconds

a. Who is the fastest runner?

Steven is the fastest runner.

> I know Steven is the fastest runner because the chart shows me that he ran 100 meters in the least number of seconds, 18 seconds.

b. Who is the slowest runner?

Sharon is the slowest runner.

> I know Sharon is the slowest runner because the chart shows me that she ran 100 meters in the most number of seconds, 24 seconds.

c. How many seconds faster did Eric run than Sharon?

$24 - 19 = 5$

Eric ran 5 seconds faster than Sharon.

> I can subtract Eric's time from Sharon's time to find how much faster Eric ran than Sharon. I can use the compensation strategy to think of subtracting $24 - 19$ as $25 - 20$ to get 5. It is much easier for me to subtract $25 - 20$ than $24 - 19$.

Lesson 1: Explore time as a continuous measurement using a stopwatch.

Name _____ Date _____

1. The table to the right shows how much time it takes each of the 5 students to run 100 meters.

Samantha	19 seconds
Melanie	22 seconds
Chester	26 seconds
Dominique	18 seconds
Louie	24 seconds

   a. Who is the fastest runner?

   b. Who is the slowest runner?

   c. How many seconds faster did Samantha run than Louie?

2. List activities at home that take about the following amounts of time to complete. If you do not have a stopwatch, you can use the strategy of counting by *1 Mississippi, 2 Mississippi, 3 Mississippi, ...*

Time	Activities at home
30 seconds	Example: Tying shoelaces
45 seconds	
60 seconds	

Lesson 1: Explore time as a continuous measurement using a stopwatch.

3. Match the analog clock with the correct digital clock.

07:05

11:00

10:15

02:50

Use a number line to answer the problems below.

1. Celina cleans her room for 42 minutes. She starts at 9:04 a.m. What time does Celina finish cleaning her room?

   I can draw a number line to help me figure out when Celina finishes cleaning her room. On the number line, I can label the first tick mark 0 and the last tick mark 60. Then I can label the hours and the 5-minute intervals.

   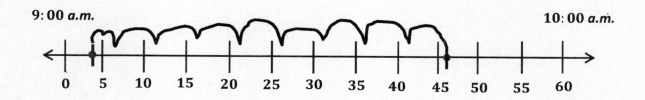

   Celina finishes cleaning her room at 9:46 a.m.

   I can plot 9:04 a.m. on the number line. Then I can count 2 minutes to 9:06 and 40 minutes by fives until 9:46. 42 minutes after 9:04 a.m. is 9:46 a.m.

2. The school orchestra puts on a concert for the school. The concert lasts 35 minutes. It ends at 1:58 p.m. What time did the concert start?

   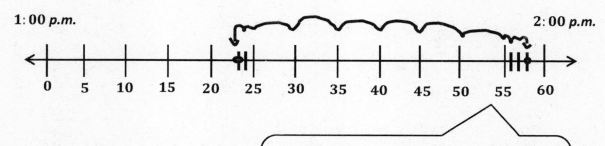

   The concert started at 1:23 p.m.

   I can plot 1:58 p.m. on the number line. Then I can count backwards from 1:58 by ones to 1:55, by fives to 1:25, and by ones to 1:23. 1:23 p.m. is 35 minutes before 1:58 p.m.

Lesson 2: Solve word problems involving time intervals within 1 hour by counting backward and forward using the number line and clock.

Name _____ Date _____

Record your homework start time on the clock in Problem 6.

Use a number line to answer Problems 1 through 4.

1. Joy's mom begins walking at 4:12 p.m. She stops at 4:43 p.m. How many minutes does she walk?

   Joy's mom walks for _____ minutes.

2. Cassie finishes softball practice at 3:52 p.m. after practicing for 30 minutes. What time did Cassie's practice start?

   Cassie's practice started at _____ p.m.

3. Jordie builds a model from 9:14 a.m. to 9:47 a.m. How many minutes does Jordie spend building his model?

   Jordie builds for _____ minutes.

4. Cara finishes reading at 2:57 p.m. She reads for a total of 46 minutes. What time did Cara start reading?

   Cara started reading at _____ p.m.

Lesson 2: Solve word problems involving time intervals within 1 hour by counting backward and forward using the number line and clock.

5. Jenna and her mom take the bus to the mall. The clocks below show when they leave their house and when they arrive at the mall. How many minutes does it take them to get to the mall?

Time when they leave home:

Time when they arrive at the mall:

6. Record your homework start time:

Record the time when you finish Problems 1–5:

How many minutes did you work on Problems 1–5?

A STORY OF UNITS – TEKS EDITION

Lesson 3 Homework Helper 3•2

Luke exercises. He stretches for 8 minutes, runs for 17 minutes, and walks for 10 minutes.

a. How many total minutes does he spend exercising?

> I can draw a strip diagram to show all the known information. I see all the parts are given, but the whole is unknown. So, I can label the whole with a question mark.

? minutes

| 8 min | 17 min | 10 min |

> I can estimate to draw the parts of my strip diagram to match the lengths of the minutes. 8 minutes is the shortest time, so I can draw it as the shortest unit. 17 minutes is the longest time, so I can draw it as the longest unit.

$8 + 17 + 10 = 35$

Luke spends a total of 35 minutes exercising.

> I can write an addition equation to find the total number of minutes Luke spends exercising. I also need to remember to write a statement that answers the question.

Lesson 3: Solve word problems involving time intervals within 1 hour by adding and subtracting on the number line.

b. Luke wants to watch a movie that starts at 1:55 p.m. It takes him 10 minutes to take a shower and 15 minutes to drive to the theater. If Luke starts exercising at 1:00 p.m., can he make it on time for the movie? Explain your reasoning.

I can draw a number line to show my reasoning. I can plot the starting time as 1:35 because I know it takes Luke 35 minutes to exercise from part (a). Then I can add 10 minutes for his shower and an additional 15 minutes for the drive to the theater.

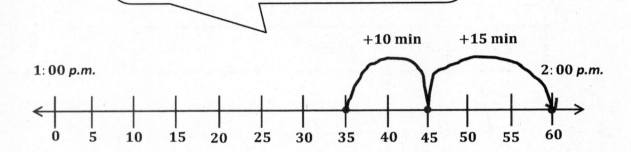

No, Luke can't make it on time for the movie. From the number line, I can see that he will be five minutes late.

I can see on the number line that Luke will be at the theater at 2:00 p.m. The movie starts at 1:55 p.m., so he'll be 5 minutes too late.

Name _____ Date _____

1. Abby spent 22 minutes working on her science project yesterday and 34 minutes working on it today. How many minutes did Abby spend working on her science project altogether? Model the problem on the number line, and write an equation to solve.

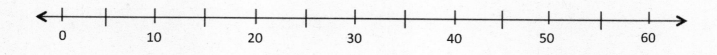

Abby spent _____ minutes working on her science project.

2. Susanna spends a total of 47 minutes working on her project. How many more minutes than Susanna does Abby spend working? Draw a number line to model the problem, and write an equation to solve.

3. Peter practices violin for a total of 55 minutes over the weekend. He practices 25 minutes on Saturday. How many minutes does he practice on Sunday?

4. a. Marcus gardens. He pulls weeds for 18 minutes, waters for 13 minutes, and plants for 16 minutes. How many total minutes does he spend gardening?

   b. Marcus wants to watch a movie that starts at 2:55 p.m. It takes 10 minutes to drive to the theater. If Marcus starts the yard work at 2:00 p.m., can he make it on time for the movie? Explain your reasoning.

5. Arelli takes a short nap after school. As she falls asleep, the clock reads 3:03 p.m. She wakes up at the time shown below. How long is Arelli's nap?

Lesson 4 Homework Helper 3•2

1. Use the chart to help you answer the following questions:

1 kilogram	100 grams	10 grams	1 gram

a. Bethany puts a marker that weighs 10 grams on a pan balance. How many 1-gram weights does she need to balance the scale?

Bethany needs ten 1-gram weights to balance the scale.

> I know that it takes ten 1-gram weights to equal 10 grams.

b. Next, Bethany puts a 100-gram bag of beans on a pan balance. How many 10-gram weights does she need to balance the scale?

Bethany needs ten 10-gram weights to balance the scale.

> I know that it takes ten 10-gram weights to equal 100 grams.

c. Bethany then puts a book that weighs 1 kilogram on a pan balance. How many 100-gram weights does she need to balance the scale?

Bethany needs ten 100-gram weights to balance the scale.

> I know that it takes ten 100-gram weights to equal 1 kilogram, or 1,000 grams.

d. What pattern do you notice in parts (a)–(c)?

I notice that to make a weight in the chart it takes ten of the lighter weight to the right in the chart. For example, to make 100 grams, it takes ten 10-gram weights, and to make 1 kilogram, or 1,000 grams, it takes ten 100-gram weights. It's just like the place value chart!

Lesson 4: Build and decompose a kilogram to reason about the size and weight of 1 kilogram, 100 grams, 10 grams, and 1 gram.

101

2. Read each digital scale. Write each weight using the word *kilogram* or *gram* for each measurement.

__153 *grams*__   __3 *kilograms*__

I can write 153 grams because I know that the letter g is used to abbreviate grams.

I can write 3 kilograms because I know that the letters kg are used to abbreviate kilograms.

Name _____ Date _____

1. Use the chart to help you answer the following questions:

1 kilogram	100 grams	10 grams	1 gram

a. Isaiah puts a 10-gram weight on a pan balance. How many 1-gram weights does he need to balance the scale?

b. Next, Isaiah puts a 100-gram weight on a pan balance. How many 10-gram weights does he need to balance the scale?

c. Isaiah then puts a kilogram weight on a pan balance. How many 100-gram weights does he need to balance the scale?

d. What pattern do you notice in Parts (a–c)?

Lesson 4: Build and decompose a kilogram to reason about the size and weight of 1 kilogram, 100 grams, 10 grams, and 1 gram.

2. Read each digital scale. Write each weight using the word *kilogram* or *gram* for each measurement.

_____ _____ _____

_____ _____ _____

1. Match each object with its approximate weight.

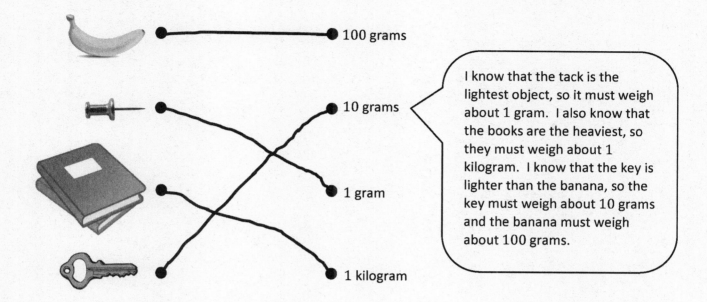

I know that the tack is the lightest object, so it must weigh about 1 gram. I also know that the books are the heaviest, so they must weigh about 1 kilogram. I know that the key is lighter than the banana, so the key must weigh about 10 grams and the banana must weigh about 100 grams.

2. Jessica weighs her dog on a digital scale. She writes 8, but she forgets to record the unit. Which unit of measurement is correct, grams or kilograms? How do you know?

   The weight of Jessica's dog needs to be recorded as 8 kilograms. Kilograms is the correct unit because 8 grams is about the same weight as 8 paperclips. It wouldn't make sense for her dog to weigh about the same as 8 paperclips.

3. Read and write the weight below. Write the word *kilogram* or *gram* with the measurement.

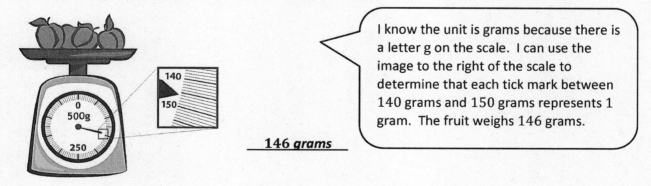

146 *grams*

I know the unit is grams because there is a letter g on the scale. I can use the image to the right of the scale to determine that each tick mark between 140 grams and 150 grams represents 1 gram. The fruit weighs 146 grams.

Lesson 5: Develop estimation strategies by reasoning about the weight in kilograms of a series of familiar objects to establish mental benchmark measures.

Name _____ Date _____

1. Match each object with its approximate weight.

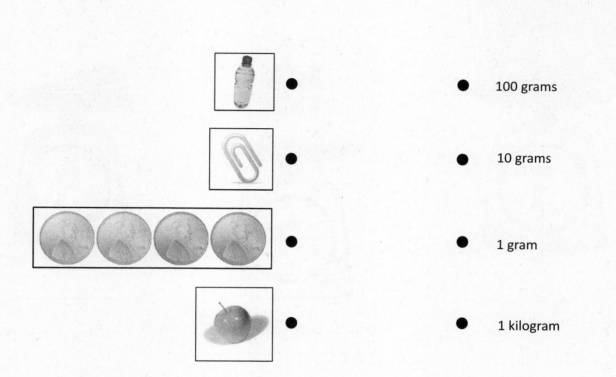

2. Alicia and Jeremy weigh a cell phone on a digital scale. They write down 113 but forget to record the unit. Which unit of measurement is correct, grams or kilograms? How do you know?

3. Read and write the weights below. Write the word *kilogram* or *gram* with the measurement.

_____

_____

_____

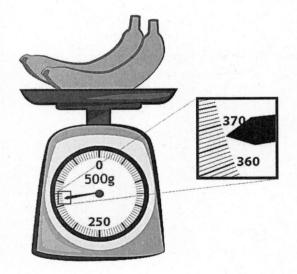

_____

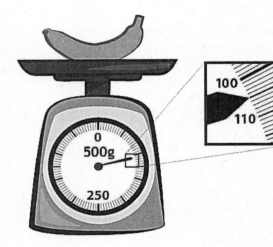

_____

The weights below show the weight of the apples in each bucket.

Bucket A
9 kg

Bucket B
7 kg

Bucket C
14 kg

> Bucket C weighs 14 kg, and Bucket B weighs 7 kg. I know that $14 - 7 = 7$, so Bucket C weighs 7 kg more.

a. The apples in Bucket __C__ are the heaviest.
b. The apples in Bucket __B__ are the lightest.
c. The apples in Bucket C are __7__ kilograms heavier than the apples in Bucket B.
d. What is the total weight of the apples in all three buckets?

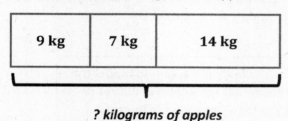

? kilograms of apples

$9 + 7 + 14 = 30$

The total weight of the apples is 30 kilograms.

> I can use a strip diagram to show the weight of each bucket of apples. Then, I can add each apple's weight to find the total weight of the apples.

e. Rebecca and her 2 sisters equally share all of the apples in Bucket A. How many kilograms of apples do they each get?

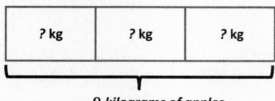

9 kilograms of apples

$9 \div 3 = 3$

Each sister gets 3 kilograms of apples.

> I know that I'm dividing 9 kilograms into 3 equal groups because 3 people are sharing the apples in Bucket A. When I know the total and the number of equal groups, I divide to find the size of each group!

f. Mason gives 3 kilograms of apples from Bucket B to his friend. He uses 2 kilograms of apples from Bucket B to make apple pies. How many kilograms of apples are left in Bucket B?

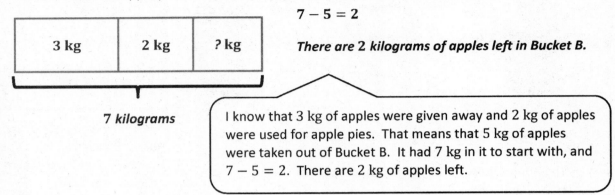

$7 - 5 = 2$

There are 2 kilograms of apples left in Bucket B.

I know that 3 kg of apples were given away and 2 kg of apples were used for apple pies. That means that 5 kg of apples were taken out of Bucket B. It had 7 kg in it to start with, and $7 - 5 = 2$. There are 2 kg of apples left.

g. Angela picks another bucket of apples, Bucket D. The apples in Bucket C are 6 kilograms heavier than the apples in Bucket D. How many kilograms of apples are in Bucket D?

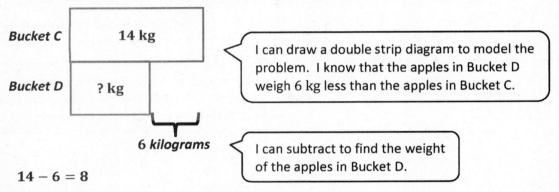

I can draw a double strip diagram to model the problem. I know that the apples in Bucket D weigh 6 kg less than the apples in Bucket C.

I can subtract to find the weight of the apples in Bucket D.

$14 - 6 = 8$

There are 8 kilograms of apples in Bucket D.

h. What is the total weight of the apples in Buckets C and D?

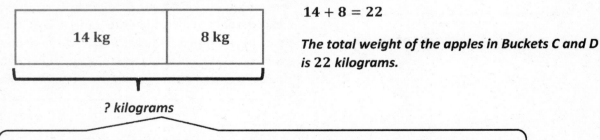

$14 + 8 = 22$

The total weight of the apples in Buckets C and D is 22 kilograms.

To find the total weight of the apples in Buckets C and D, I need to add. I know that $14 + 8 = 22$, so the total weight of the apples in Buckets C and D is 22 kilograms.

Name _____ Date _____

1. The weights of 3 fruit baskets are shown below.

Basket A
12 kg

Basket B
8 kg

Basket C
16 kg

   a. Basket _____ is the heaviest.

   b. Basket _____ is the lightest.

   c. Basket A is _____ kilograms heavier than Basket B.

   d. What is the total weight of all three baskets?

2. Each journal weighs about 280 grams. What is total weight of 3 journals?

3. Ms. Rios buys 453 grams of strawberries. She has 23 grams left after making smoothies. How many grams of strawberries did she use?

Lesson 6: Solve one-step word problems involving metric weights within 100 and estimate to reason about solutions.

4. A load of watermelons is 57 kilograms heavier than a load of squash. The load of squash weighs 34 kilograms.

   a. How much does the load of watermelons weigh?

   b. How much do the watermelons and the squash weigh in total?

5. Jennifer's grandmother buys carrots at the farm stand. She and her 3 grandchildren equally share the carrots. The total weight of the carrots she buys is shown below.

   a. How many kilograms of carrots will Jennifer get?

   b. Jennifer uses 2 kilograms of carrots to bake muffins. How many kilograms of carrots does she have left?

1. Ben makes 4 batches of cookies for the bake sale. He uses 5 milliliters of vanilla for each batch. How many milliliters of vanilla does he use in all?

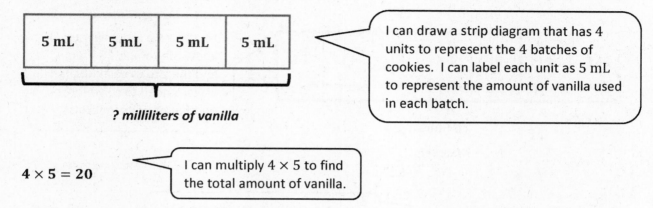

$4 \times 5 = 20$

Ben uses 20 milliliters of vanilla.

2. Mrs. Gillette pours 3 glasses of juice for her children. Each glass holds 321 milliliters of juice. How much juice does Mrs. Gillette pour in all?

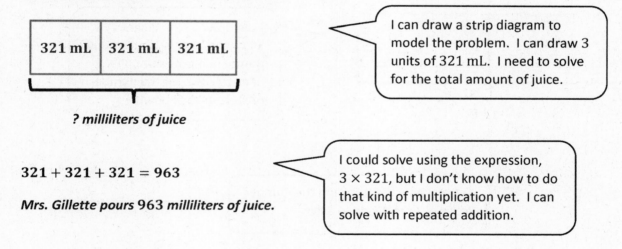

$321 + 321 + 321 = 963$

Mrs. Gillette pours 963 milliliters of juice.

3. Gabby uses a 4-liter bucket to give her pony water. How many buckets of water will Gabby need in order to give her pony 28 liters of water?

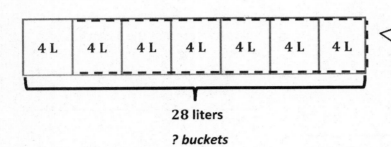

> I can draw a strip diagram. I know the total is 28 liters and the size of each unit is 4 liters. I need to solve for the number of units (buckets).

$28 \div 4 = 7$

> Since I know the total and the size of each unit, I can divide to solve.

Gabby needs 7 buckets of water.

4. Elijah makes 12 liters of punch for his birthday party. He pours the punch equally into 4 bowls. How many liters of punch are in each bowl?

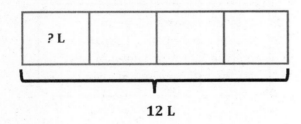

> I can draw a strip diagram. I know the total is 12 liters and there are 4 bowls or units. I need to solve for the number of liters in each bowl.

$12 \div 4 = 3$

> Since I know the total and the number of units, I can divide to solve.

Elijah pours **3** *liters of punch into each bowl.*

> I can divide to solve Problems 3 and 4, but the unknowns in each problem are different. In Problem 3, I solved for the number of groups/units. In Problem 4, I solved for the size of each group/unit.

Name _____   Date _____

1. Find containers at home that have a capacity of about 1 liter. Use the labels on containers to help you identify them.

    a.

Name of Container
Example: Carton of orange juice

    b. Sketch the containers. How do their sizes and shapes compare?

2. The doctor prescribes Mrs. Larson 5 milliliters of medicine each day for 3 days. How many milliliters of medicine will she take altogether?

3. Mrs. Goldstein pours 3 juice boxes into a bowl to make punch. Each juice box holds 236 milliliters. How much juice does Mrs. Goldstein pour into the bowl?

4. Daniel's fish tank holds 24 liters of water. He uses a 4-liter bucket to fill the tank. How many buckets of water are needed to fill the tank?

5. Sheila buys 15 liters of paint to paint her house. She pours the paint equally into 3 buckets. How many liters of paint are in each bucket?

1. Estimate the amount of liquid in each container to the nearest liter.

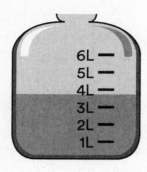

The liquid in this container is between 3 liters and 4 liters. Since it is more than halfway to the next liter, 4 liters, I can estimate that there are about 4 liters of liquid.

4 liters

The liquid in this container is at exactly 5 liters.

5 liters

The liquid in this container is between 3 liters and 4 liters. Since it is less than halfway to the next liter, 4 liters, I can estimate that there are about 3 liters of liquid.

3 liters

Lesson 8: Estimate and measure liquid volume in liters and milliliters using the vertical number line.

A STORY OF UNITS – TEKS EDITION

Lesson 8 Homework Helper 3•2

2. Manny is comparing the capacity of buckets that he uses to water his vegetable garden. Use the chart to answer the questions.

Bucket	Capacity in Liters
Bucket 1	17
Bucket 2	12
Bucket 3	23

   a. Label the number line to show the capacity of each bucket. Bucket 2 has been done for you.

   [Vertical number line with 30 L at top, 20 L, and 10 L marked. Bucket 3 is labeled at 23 L, Bucket 1 at 17 L, and Bucket 2 at 12 L.]

   > I can use the tick marks to help me locate the correct place on the number line for each bucket. I can label Bucket 1 at 17 liters and Bucket 3 at 23 liters.

   b. Which bucket has the greatest capacity?

   Bucket 3 has the greatest capacity.

   > I can use the vertical number line to help me answer both of these questions. I can see that the point I plotted for Bucket 3 is higher up the number line than the others, so it has a larger capacity than the others. I also see that the point I plotted for Bucket 2 is lowest on the number line, so it has the smallest capacity.

   c. Which bucket has the smallest capacity?

   Bucket 2 has the smallest capacity.

   d. Which bucket has a capacity of about 10 liters?

   Bucket 2 has a capacity of about 10 liters.

   > I notice that Bucket 2 is closest to 10 liters, so it has a capacity of about 10 liters.

   e. Use the number line to find how many more liters Bucket 3 holds than Bucket 2.

   Bucket 3 holds 11 more liters than Bucket 2.

   > To solve this problem, I can count up on the number line from Bucket 2 to Bucket 3. I'll start at 12 liters because that is the capacity of Bucket 2. I count up 8 tick marks to 20 liters, and then I count 3 more tick marks to 23, which is the capacity of Bucket 3. I know that $8 + 3 = 11$, so Bucket 3 holds 11 more liters than Bucket 2.

Lesson 8: Estimate and measure liquid volume in liters and milliliters using the vertical number line.

Name _____ Date _____

1. How much liquid is in each container?

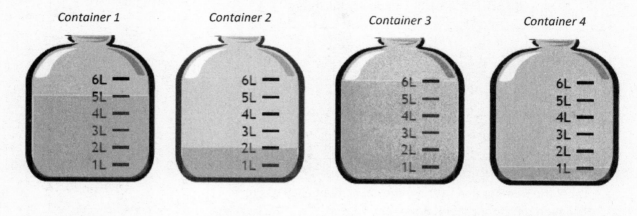

2. Jon pours the contents of Container 1 and Container 3 above into an empty bucket. How much liquid is in the bucket after he pours the liquid?

3. Estimate the amount of liquid in each container to the nearest liter.

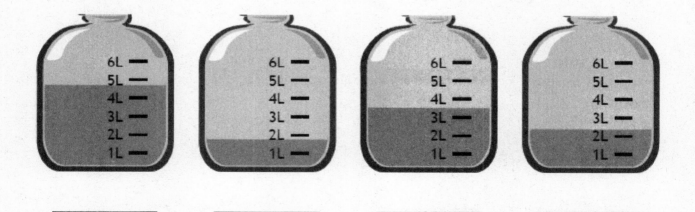

4. Kristen is comparing the capacity of gas tanks in different size cars. Use the chart below to answer the questions.

Size of Car	Capacity in Liters
Large	74
Medium	57
Small	42

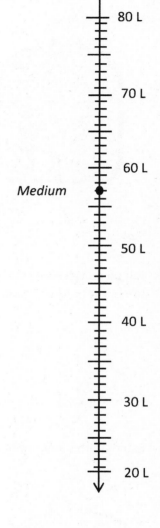

a. Label the number line to show the capacity of each gas tank. The medium car has been done for you.

b. Which car's gas tank has the greatest capacity?

c. Which car's gas tank has the smallest capacity?

d. Kristen's car has a gas tank capacity of about 60 liters. Which car from the chart has about the same capacity as Kristen's car?

e. Use the number line to find how many more liters the large car's tank holds than the small car's tank.

1. Together the weight of a banana and an apple is 291 grams. The banana weighs 136 grams. How much does the apple weigh?

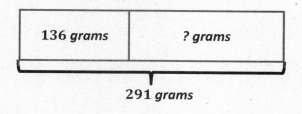

I can draw a strip diagram to model the problem. The total is 291 grams, and one part—the weight of the banana—is 136 grams. I can subtract to find the other part, the weight of the apple.

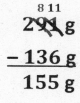

I can use the standard algorithm to subtract. I can unbundle 1 ten to make 10 ones. Now there are 2 hundreds, 8 tens, and 11 ones.

The apple weighs 155 grams.

2. Sandy uses a total of 21 liters of water to water her flowerbeds. She uses 3 liters of water for each flowerbed. How many flowerbeds does Sandy water?

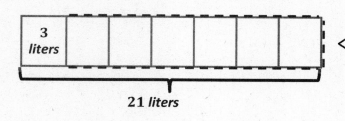

I can draw a strip diagram to model the problem. The total is 21 liters, and each unit represents the amount of water Sandy uses for each flowerbed, 3 liters. I can see that the unknown is the number of units (groups).

$21 \div 3 = 7$

I can divide to find the total number of units, which represents the number of flowerbeds.

Sandy waters 7 flowerbeds.

Now that I know the answer, I can draw the rest of the units in my strip diagram, to show a total of 7 units.

Name _____ Date _____

1. Karina goes on a hike. She brings a notebook, a pencil, and a camera. The weight of each item is shown in the chart. What is the total weight of all three items?

Item	Weight
Notebook	312 g
Pencil	10 g
Camera	365 g

The total weight is _____ grams.

2. Together a horse and its rider weigh 729 kilograms. The horse weighs 625 kilograms. How much does the rider weigh?

The rider weighs _____ kilograms.

3. Theresa's soccer team fills up 6 water coolers before the game. Each water cooler holds 9 liters of water. How many liters of water do they fill?

4. Dwight purchased 48 kilograms of fertilizer for his vegetable garden. He needs 6 kilograms of fertilizer for each bed of vegetables. How many beds of vegetables can he fertilize?

5. Nancy bakes 7 cakes for the school bake sale. Each cake requires 5 milliliters of oil. How many milliliters of oil does she use?

A STORY OF UNITS – TEKS EDITION

Lesson 10 Homework Helper   3•2

1. Rewrite the number, including a comma where appropriate:

   15024   **15,024**

   > I place a comma between the hundreds place and the thousands place.

2. Label the missing unit on the place value chart. Write the number in standard form.

hundred thousands	ten thousands	thousands	hundreds	tens	ones
	• • • •	• • • • • • •		• • •	• • • • • •

   > The place value chart shows the value of each digit. The dots show the number of each unit. I see there are 4 ten thousands, 7 thousands, 3 tens, and 6 ones.

   Standard form: **47,036**

3. 5 ten thousands 9 thousands 6 tens 2 hundreds 4 ones

   59,264

   > When I look at the unit form of the number, I need to make sure the units are in order. Here, the tens and hundreds are out of order. When I write the number in standard form, I have to be careful to put the digits in the correct order and to place the comma between the hundreds and the thousands place.

Lesson 10: Name numbers up to 100,000 by building understanding of the place value chart and placement of commas for naming base thousand units.

3. Solve each expression. Record your answer in unit form and in standard form.

Expression	Unit Form	Standard Form
7 hundreds + 3 hundreds	__10__ hundreds = __1__ thousands	1,000

(I can add 7 hundreds + 3 hundreds = 10 hundreds.)

(10 hundreds are equal to 1 thousand, or 1,000.)

4. Represent each addend with place value disks in the place value chart. Show the bundling of smaller units to make larger units. Write the sum in standard form.

4 thousands + 13 hundreds = __5,300__

hundred thousands	ten thousands	thousands	hundreds	tens	ones
		●●●● ●	●●●●● ●●●●● ●●●		

After drawing 4 thousands disks and 13 hundreds disks, I see that 10 hundreds can be bundled as 1 thousand. Now my drawing shows 5 thousands 3 hundreds, or 5,300.

Name _____ Date _____

1. Rewrite the numbers including commas where appropriate:

   a. 7852 _____   b. 97852 _____   c. 10000 _____

2. Label the missing units on the place value chart. Write the number in standard form.

   a.

hundred thousands		thousands		tens	ones
	••••	••••• •••	•••••		••••• •••

   Standard form: _____

   b.

hundred thousands	ten thousands			tens	ones
	••••• ••	••		•••	•

   Standard form: _____

3. Write the numbers in standard form.

   a. 8 ten thousands 6 hundreds 9 thousands 4 tens 7 ones = _____

   b. 4 ten thousands 5 thousands 9 tens 6 ones 2 hundreds = _____

4. Solve each expression. Record your answer in unit form and in standard form.

Expression	Unit Form	Standard Form
1 ten + 9 tens	____ tens = _____ hundreds	
7 hundreds + 3 hundreds	_____ hundreds = _____ thousands	
8 thousands + 2 thousands	_____ thousands = _____ ten thousands	
6 ten thousands + 4 ten thousands	____ ten thousands = ___ hundred thousands	
50 thousands + 50 thousands	_____ thousands = 1 _____	

5. Represent each addend with place value disks in the place value chart. Show the bundling of smaller units to make larger units. Write the sum in standard form.

   a. 4 thousands + 12 hundreds = _____

hundred thousands	ten thousands	thousands	hundreds	tens	ones

b. 4 ten thousands + 12 thousands = _____

hundred thousands	ten thousands	thousands	hundreds	tens	ones

c. 9 thousands + 10 hundreds = _____

hundred thousands	ten thousands	thousands	hundreds	tens	ones

6. Jayden has 50 thousands disks. He wants to trade them for some ten thousands disks. How many ten thousands disks are equal to 50 thousands disks?

Lesson 11 Homework Helper 3•2

1. a. On the place value chart below, label the units, and represent the number 23,146

hundred thousands	ten thousands	thousands	hundreds	tens	ones
	• •	• • •	•	• • • •	• • • • • •

b. Write the number in word form. **twenty-three thousand, one hundred forty-six**

> To write the word form, I read 23,146 to myself. Then I write the words that I say. I write a comma to separate the thousands from the hundreds, tens, and ones.

c. Write the number in expanded form. **20,000 + 3,000 + 100 + 40 + 6**

> To write a number in expanded form, I write the value of each digit in 23,146 as an addition equation. For example, the 2 has a value of 2 ten thousands, which I write in standard form as 20,000. 23,146 = 20,000 + 3,000 + 100 + 40 + 6.

d. Write the number in expanded notation. **(2 × 10,000) + (3 × 1,000) + (1 × 100) + (4 × 10) + (6 × 1)**

> To write a number, I write the value of each digit in 23,146 as a multiplication expression and then add those expressions together. For example, there are 2 copies of 10,000 and 3 copies of 1,000, so I write 2 × 10,000 and 3 × 1,000. I write all the other digits' values by telling how many copies and then writing the value of the place value unit. Once that is done, I add them all together.

Lesson 11: Read and write numbers up to 100,000 using base ten numerals, number names, expanded form, and expanded notation.

2. Paloma wrote 90,000 + 3,000 + 800 + 20 + 6 = 903,826.

   Explain her error and write the number correctly.

   93,826

   > I notice the expanded form says 90 thousand. It looks like Paloma thought she should write 90 instead of just a 9 in the ten thousands place. 90,000 means 9 ten thousands. She should have written the number as 93,826.

Name _____ Date _____

1. a. On the place value chart below, label the units and write the number 90,745.

   b. Write the number in word form.

   c. Write the number in expanded form. _____

   d. Write the number in expanded notation. _____

2. a. On the place value chart below, label the units and write the number 97,405.

   b. Write the number in word form.

   c. Write the number in expanded form. _____

   d. Write the number in expanded notation. _____

Lesson 11: Read and write numbers up to 100,000 using base ten numerals, number names, expanded form, and expanded notation.

Name _____ Date _____

3. Complete the following chart:

Standard Form	Expanded Form	Expanded Notation
7,917		(7 × 1,000) + (9 × 100) + (1 × 10) + (7 × 1)
	70,000 + 2,000 + 300 + 9	
83,257		
		(4 × 10,000) + (2 × 1,000) + (6 × 10) + (3 × 1)
42,603		

4. Jasper wrote thirty-two thousand, five hundred eight as 302,508. Explain his error and write the number correctly.

1. Label the units in the place value chart. Draw place value disks to represent each number in the place value chart. Use <, >, or = to compare the two numbers. Write the correct symbol in the circle.

> I record the comparison symbol for *less than*.

53,721 $\bigcirc$ < $\bigcirc$ 57,021

hundred thousands	ten thousands	thousands	hundreds	tens	ones
	● ● ● ● ●	● ● ●	● ● ● ● ● ● ●	● ●	●
	● ● ● ● ●	● ● ● ● ● ● ●		● ●	●

> I record the value of each digit using dots. I place 53,721 in one row and 57,021 in another row of the place value chart. I can clearly see that the value of the largest unit is the same for both numbers. I compare the values of the digits for the next largest unit—thousands. 3 thousands is less than 7 thousands. 53,721 is less than 57,021.

2. Compare the two numbers by using the symbols <, >, or =. Write the correct symbol in the circle.

thirty-six thousand, eight hundred-five $\bigcirc$ > $\bigcirc$ 30,000 + 6,000 + 700 + 80 + 5

> It helps me to solve if I write both numbers in standard form.

36,805 $\bigcirc$ > $\bigcirc$ 36,785

> The value of the largest unit is the same, and the value of the next largest unit is the same. I compare the third largest unit—the hundreds. Eight hundreds is greater than seven hundreds. So, 36,805 is greater than 36,785. I record the comparison symbol for *greater than* to complete my answer.

Lesson 12: Compare numbers based on the meaning of the digits using <, >, or = to record the comparison.

3. Four people are competing in a game. At the end of the game, Josh has 22,025 points, Valerie has 21,725 points, Kim has 23,000 points, and Carlos has 22,725 points. Who has the most points at the end of the game?

> Listing the amounts of money in a place value chart helps me to see the values in each unit.

	hundred thousands	ten thousands	thousands	hundreds	tens	ones
Josh		2	2	0	2	5
Valerie		2	1	7	2	5
Kim		2	3	0	0	0
Carlos		2	2	7	2	5

> I notice that all four numbers have 2 ten thousands. 22,025 and 22,725 both have 2 thousands, but 22,725 has more hundreds. I also notice that 21,725 has only 1 thousand, but 23,000 has 3 thousands.

23,000 > 22,725 > 22,025 > 21,725

Kim has the most points at the end of the game.

Name _____ Date _____

1. Label the units in the place value chart. Draw place value disks to represent each number in the place value chart. Use <, >, or = to compare the two numbers. Write the correct symbol in the circle.

   a. 1,823 ◯ 10,023

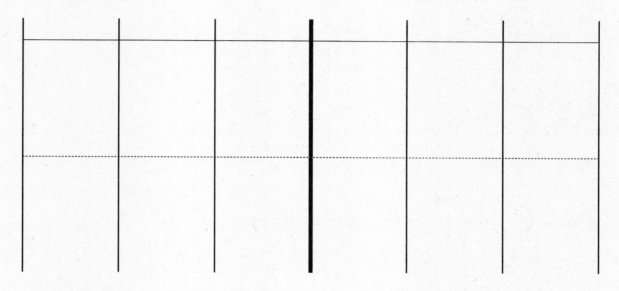

   b. 83,042 ◯ 81,342

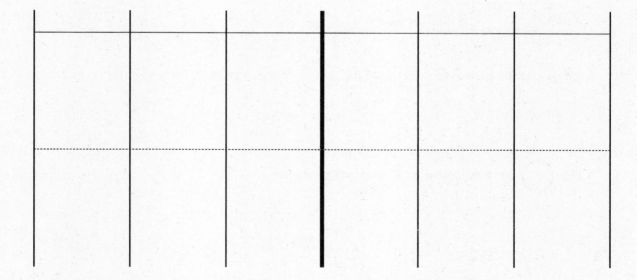

Lesson 12: Compare numbers based on the meaning of the digits using <, >, or = to record the comparison.

2. Compare the two numbers by using the symbols <, >, or =. Write the correct symbol in the circle.

   a. 20,084 ◯ 2,584

   b. 73,000 ◯ 71,499

   c. 45,622 ◯ 45,266

   d. 60,352 ◯ 6 ten thousands 3 hundreds 5 tens 2 ones

   e. (9 × 10,000) + (6 × 100) + (3 × 1) ◯ 90,000 + 600 + 3

   f. 3 ten thousands 2 thousands 7 tens 4 ones ◯ fifteen thousand, forty-seven

   g. 4,125 ◯ forty thousand, one hundred twenty-five

3. A group of friends will participate in a 5-kilometer walk to raise money for a local charity. The chart shows the total number of steps they each walked during the week before the 5-kilometer walk.

List the numbers of steps in order from greatest to least. Then, name the person who took the fewest steps that week.

Name	Number of Steps
Camila	43,098
Emily	42,199
Jasmine	43,100
Lanelle	44,005
Mia	42,250

4. The chart lists all the Texas counties where the number of people under the age of 18 was between 50,000 and 100,000 during the 2010 census.

   a. List the names of the counties in order from least to greatest according to the number of people under the age of 18.

County in Texas	People Under the Age of 18 in 2010
Bell	88,117
Brazoria	86,985
Galveston	74,167
Jefferson	60,398
Lubbock	67,862
McLennan	59,745
Nueces	88,255
Smith	53,796
Webb	88,158

   b. Kim made a list of the counties in order from least to greatest according to the total population in 2010 is as follows: Smith, McLennan, Webb, Jefferson, Lubbock, Galveston, Bell, Brazoria, Nueces. Does Kim's list match the order in Part (a)? Why or why not?

1. Complete the chart.

   "I measured the width of a picture frame. It was 24 centimeters wide."

Object	Measurement (in cm)	The object measures between (which two tens)...	Length rounded to the nearest 10 cm
Width of picture frame	24 cm	__20__ and __30__ cm	20 cm

   "I can use a vertical number line to help me round 24 cm to the nearest 10 cm."

   "The endpoints on my vertical number line help me know which two tens the width of the picture frame is in between."

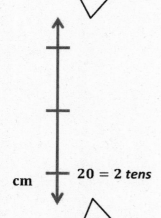

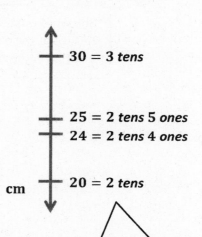

There are 2 tens in 24, so I can label this endpoint as 2 tens or 20.

One more ten than 2 tens is 3 tens, so I can label the other endpoint as 3 tens or 30. Halfway between 2 tens and 3 tens is 2 tens 5 ones. I can label the halfway point as 2 tens 5 ones or 25.

I can plot 24 or 2 tens 4 ones on the vertical number line. I can easily see that 24 is less than halfway between 2 tens and 3 tens. That means that 24 cm rounded to the nearest 10 cm is 20 cm.

Lesson 13: Round two-digit measurements to the nearest ten on the vertical number line.

2. Measure the liquid in the beaker to the nearest 10 milliliters.

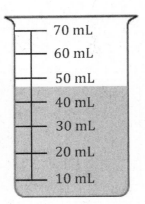

I can use the beaker to help me round the amount of liquid to the nearest 10 mL. I can see that the liquid is between 40 (4 tens) and 50 (5 tens). I can also see that the liquid is more than halfway between 4 tens and 5 tens. That means that the amount of liquid rounds up to the next ten milliliters, 50 mL.

There are about __50__ milliliters of liquid in the beaker.

The word *about* tells me that this is not the exact amount of liquid in the beaker.

Name _____ Date _____

1. Complete the chart. Choose objects, and use a ruler or meter stick to complete the last two on your own.

Object	Measurement (in cm)	The object measures between (which two tens)...	Length rounded to the nearest 10 cm
Length of desk	66 cm	_____ and _____ cm	
Width of desk	48 cm	_____ and _____ cm	
Width of door	81 cm	_____ and _____ cm	
		_____ and _____ cm	
		_____ and _____ cm	

2. Gym class ends at 10:27 a.m. Round the time to the nearest 10 minutes.

Gym class ends at about _____ a.m.

3. Measure the liquid in the beaker to the nearest 10 milliliters.

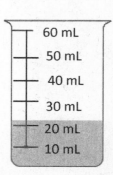

There are about _____ milliliters in the beaker.

Lesson 13: Round two-digit measurements to the nearest ten on the vertical number line.

4. The scale at the zoo shows a leopard's weight. Round her weight to the nearest 10 kilograms.

The Leopard's weight is _____ kilograms.

The Leopard weighs about _____ kilograms.

5. A zookeeper weighs a chimp. Round the chimp's weight to the nearest 10 kilograms.

The chimp's weight is _____ kilograms.

The chimp weighs about _____ kilograms.

1. Round to the nearest ten. Draw a number line to model your thinking.

   a. 52 ≈ __50__

   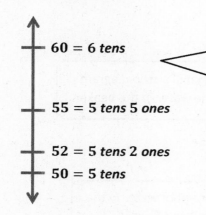

   I can draw a vertical number line with endpoints of 50 and 60 and a halfway point of 55. When I plot 52 on the vertical number line, I can see that it is less than halfway between 50 and 60. So 52 rounded to the nearest ten is 50.

   b. 152 ≈ __150__

   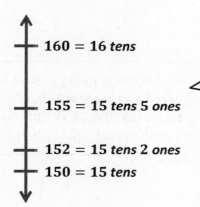

   I can draw a vertical number line with endpoints of 150 and 160 and a halfway point of 155. When I plot 152 on the vertical number line, I can see that it is less than halfway between 150 and 160. So 152 rounded to the nearest ten is 150.

   Look, my vertical number lines for parts (a) and (b) are almost the same! The only difference is that all the numbers in part (b) are 100 more than the numbers in part (a).

2. Amelia pours 63 mL of water into a beaker. Madison pours 56 mL of water into Amelia's beaker. Round the total amount of water in the beaker to the nearest 10 milliliters. Model your thinking using a number line.

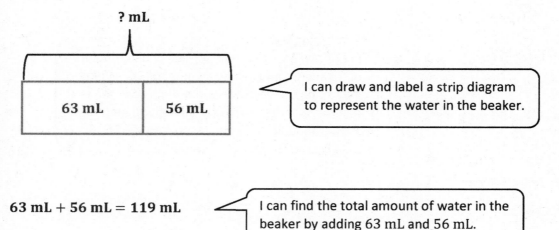

$63 \text{ mL} + 56 \text{ mL} = 119 \text{ mL}$

I can find the total amount of water in the beaker by adding 63 mL and 56 mL.

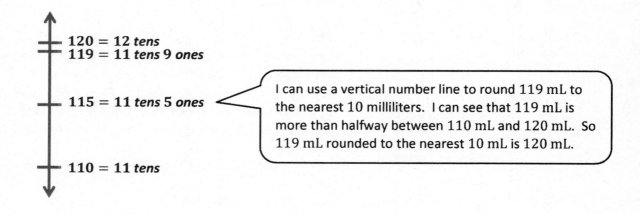

I can use a vertical number line to round 119 mL to the nearest 10 milliliters. I can see that 119 mL is more than halfway between 110 mL and 120 mL. So 119 mL rounded to the nearest 10 mL is 120 mL.

There are about 120 mL of water in the beaker.

Name _____ Date _____

1. Round to the nearest ten. Use the number line to model your thinking.

a. 43 ≈ _____ 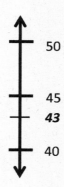	b. 48 ≈ _____ 
c. 73 ≈ _____ 	d. 173 ≈ _____ 
e. 189 ≈ _____ 	f. 194 ≈ _____ 

Lesson 14: Round two- and three-digit numbers to the nearest ten on the vertical number line.

2. Round the weight of each item to the nearest 10 grams. Draw number lines to model your thinking.

Item	Number Line	Round to the nearest 10 grams
Cereal bar: 45 grams		
Loaf of bread: 673 grams		

3. The Garden Club plants rows of carrots in the garden. One seed packet weighs 28 grams. Round the total weight of 2 seed packets to the nearest 10 grams. Model your thinking using a number line.

1. Round to the nearest hundred. Draw a number line to model your thinking.

   a. $234 \approx$ __200__

   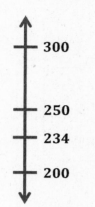

   I can draw a vertical number line with endpoints of 200 and 300 and a halfway point of 250. When I plot 234 on the vertical number line, I can see that it is less than halfway between 200 and 300. So 234 rounded to the nearest hundred is 200.

   b. $1{,}234 \approx$ __1,200__

   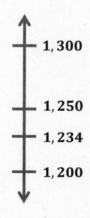

   I can draw a vertical number line with endpoints of 1,200 and 1,300 and a halfway point of 1,250. When I plot 1,234 on the vertical number line, I can see that it is less than halfway between 1,200 and 1,300. So 1,234 rounded to the nearest hundred is 1,200.

   Look, my vertical number lines for parts (a) and (b) are almost the same! The only difference is that all the numbers in part (b) are 1,000 more than the numbers in part (a).

Lesson 15: Round to the nearest hundred on the vertical number line.

2. There are 1,365 students at Park Street School. Kate and Sam round the number of students to the nearest hundred. Kate says it is one thousand, four hundred. Sam says it is 14 hundreds. Who is correct? Explain your thinking.

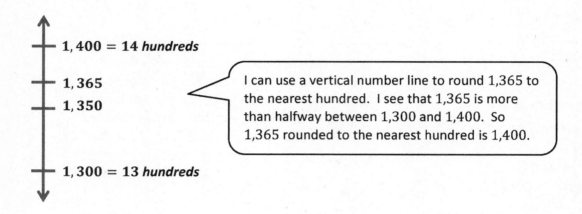

I can use a vertical number line to round 1,365 to the nearest hundred. I see that 1,365 is more than halfway between 1,300 and 1,400. So 1,365 rounded to the nearest hundred is 1,400.

Kate and Sam are both right. 1,365 rounded to the nearest hundred is 1,400. 1,400 in unit form is 14 hundreds.

Name _____ Date _____

1. Round to the nearest hundred. Use the number line to model your thinking.

a. 156 ≈ _____

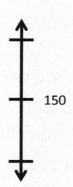 150

b. 342 ≈ _____

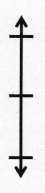

c. 260 ≈ _____

d. 1,260 ≈ _____

e. 1,685 ≈ _____

f. 1,804 ≈ _____

2. Complete the chart.

a. Luis has 217 baseball cards. Round the number of cards Luis has to the nearest hundred.	
b. There were 462 people sitting in the audience. Round the number of people to the nearest hundred.	
c. A bottle of juice holds 386 milliliters. Round the capacity to the nearest 100 milliliters.	
d. A book weighs 727 grams. Round the weight to the nearest 100 grams.	
e. Joanie's parents spent $1,260 on two plane tickets. Round the total to the nearest $100.	

3. Circle the numbers that round to 400 when rounding to the nearest hundred.

   368   342   420   492   449   464

4. There are 1,525 pages in a book. Julia and Kim round the number of pages to the nearest hundred. Julia says it is one thousand, five hundred. Kim says it is 15 hundreds. Who is correct? Explain your thinking.

A STORY OF UNITS – TEKS EDITION

Lesson 16 Homework Helper 3•2

1. Round to the nearest ten thousand. Use the number line to model your thinking.

    41,384 ≈ **40,000**

    ```
    ↑
    — 50,000 = 5 ten thousands

    — 45,000 = 4 ten thousands 5 thousands

    — 41,384
    — 40,000 = 4 ten thousands
    ↓
    ```

    > I can draw a vertical number line with endpoints 40,000 and 50,000 and a halfway point of 45,000. When I plot 41,384 on the number line, I can see that it is less than halfway between 40,000 and 50,000. 41,384 rounded to the nearest ten thousand is 40,000.

2. There are 8,615 students who go to the College of William and Mary. About how many students go to the college? Round to the nearest thousand. Draw a number line to model your thinking.

    8,615 ≈ **9,000**

    ```
    ↑
    — 9,000 = 9 thousands

    — 8,615
    — 8,500 = 8 thousands 5 hundreds

    — 8,000 = 8 thousands
    ↓
    ```

    > I can draw a vertical number line to round 8,615. When I plot 8,615 on the number line, I can see that it is more than halfway between 8,000 and 9,000. 8,615 rounded to the nearest thousand is 9,000.

    About 9,000 students go to the College of William and Mary.

3. Round 4,283:

    - To the nearest thousand **4,000**

    - To the nearest hundred **4,300**

    - To the nearest ten **4,280**

    > I can use the vertical number line to help me round this number to different places. When I round to a new place, my endpoints change.

Lesson 16: Round four- and five-digit numbers using the vertical number line.

Name _____   Date _____

1. Round to the place value indicate in the box. Use the number line to model your thinking.

   a. Nearest thousand

   1,242 ≈ _____

   b. Nearest thousand

   2,976 ≈ _____

   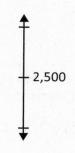

   c. Nearest hundred

   923 ≈ _____

   d. Nearest ten thousand

   15,819 ≈ _____

   e. Nearest ten thousand

   37,581 ≈ _____

   f. Nearest ten thousand

   94,523 ≈ _____

A STORY OF UNITS – TEKS EDITION Lesson 16 Homework 3•2

2. Circle the numbers that round to 6,000 when rounding to the nearest thousand.

   6,499   5,499   6,729   6,192   5,932   6,004

3. There are 4,519 fans at the Dogwood Baseball Stadium for Saturday afternoon's game. About how many fans attended the game? Round to the nearest thousand. Draw a number line to model your thinking.

4. Round 4,358:
   - To the nearest thousand _____
   - To the nearest hundred _____
   - To the nearest ten _____

   Use a vertical number line, if needed.

5. Round 73,908 to the nearest ten thousand. Use a vertical number line, if needed.

Lesson 17 Homework Helper   3•2

1. Find the sums below. Choose mental math or the algorithm.

   a. 69 cm + 7 cm = **76 cm**

         / \
        1   6
       70

   > I can use mental math to solve this problem. I broke apart the 7 as 1 and 6. Then I solved the equation as 70 cm + 6 cm = 76 cm.

   > For this problem, the standard algorithm is a more strategic tool to use.

   b. 59 kg + 76 kg

   $$\begin{array}{r}59 \text{ kg} \\ +\,76 \text{ kg} \\ \hline 5\end{array}$$
   (small 1 under the tens line)

   $$\begin{array}{r}59 \text{ kg} \\ +\,76 \text{ kg} \\ \hline 135 \text{ kg}\end{array}$$
   (small 1 under the tens line)

   > 9 ones plus 6 ones is 15 ones. I can rename 15 ones as 1 ten and 5 ones. I can record this by writing the 1 so that it crosses the line under the tens in the tens place, and the 5 below the line in the ones column. This way I write 15, rather than 5 and 1 as separate numbers.

   > 5 tens plus 7 tens plus 1 ten equals 13 tens. So, 59 kg + 76 kg = 135 kg.

Lesson 17: Add measurements using the standard algorithm to compose larger units once.

2. Mrs. Alvarez's plant grew 23 centimeters in one week. The next week it grew 6 centimeters more than the previous week. What is the total number of centimeters the plant grew in 2 weeks?

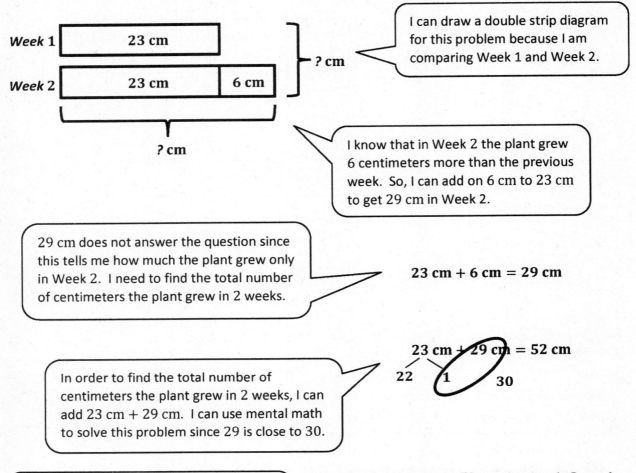

I can draw a double strip diagram for this problem because I am comparing Week 1 and Week 2.

I know that in Week 2 the plant grew 6 centimeters more than the previous week. So, I can add on 6 cm to 23 cm to get 29 cm in Week 2.

29 cm does not answer the question since this tells me how much the plant grew only in Week 2. I need to find the total number of centimeters the plant grew in 2 weeks.

23 cm + 6 cm = 29 cm

In order to find the total number of centimeters the plant grew in 2 weeks, I can add 23 cm + 29 cm. I can use mental math to solve this problem since 29 is close to 30.

23 cm + 29 cm = 52 cm
22   1   30

Now I can write a statement that answers the question. This helps me check my work to see if my answer is reasonable.

The plant grew 52 centimeters in 2 weeks.

Name _____ Date _____

1. Find the sums below. Choose mental math or the algorithm.

   a. 75 cm + 7 cm

   c. 362 mL + 229 mL

   e. 451 mL + 339 mL

   b. 39 kg + 56 kg

   d. 283 g + 92 g

   f. 149 L + 331 L

2. The liquid volume of five drinks is shown below.

Drink	Liquid Volume
Apple juice	125 mL
Milk	236 mL
Water	248 mL
Orange juice	174 mL
Fruit punch	208 mL

   a. Jen drinks the apple juice and the water. How many milliliters does she drink in all?

   Jen drinks _____ mL.

   b. Kevin drinks the milk and the fruit punch. How many milliliters does he drink in all?

Lesson 17: Add measurements using the standard algorithm to compose larger units once.

3. There are 75 students in Grade 3. There are 44 more students in Grade 4 than in Grade 3. How many students are in Grade 4?

4. Mr. Green's sunflower grew 29 centimeters in one week. The next week it grew 5 centimeters more than the previous week. What is the total number of centimeters the sunflower grew in 2 weeks?

5. Kylie records the weights of 3 objects as shown below. Which 2 objects can she put on a pan balance to equal the weight of a 460 gram bag? Show how you know.

Paperback Book	Banana	Bar of Soap
343 grams	108 grams	117 grams

1. Find the sums.

   a. 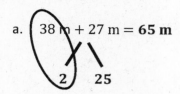 38 m + 27 m = **65 m**

   (break apart: 2 and 25)

   > I can use mental math to solve this problem. I can break apart 27 as 2 and 25. Then I can solve 40 m + 25 m, which is 65 m.

   b. 358 kg + 167 kg

   > I can use the standard algorithm to solve this problem. I can line the numbers up vertically and add.

   ```
     385 kg           385 kg           385 kg
   + 167 kg         + 167 kg         + 167 kg
   ───1───          ──11───          ──11───
        2               52             552 kg
   ```

   > 5 ones plus 7 ones is 12 ones. I can rename 12 ones as 1 ten 2 ones.

   > 8 tens plus 6 tens is 14 tens. Plus 1 more ten is 15 tens. I can rename 15 tens as 1 hundred 5 tens.

   > 3 hundreds plus 1 hundred is 4 hundreds. Plus 1 more hundred is 5 hundreds. The sum is 552 kg.

Lesson 18: Add measurements using the standard algorithm to compose larger units twice.

2. Matthew reads for 58 more minutes in March than in April. He reads for 378 minutes in April. Use a strip diagram to find the total minutes Matthew reads in March and April.

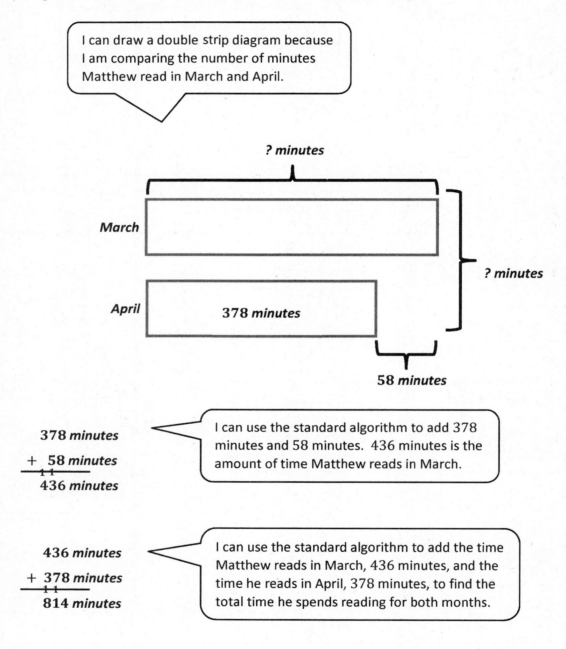

I can draw a double strip diagram because I am comparing the number of minutes Matthew read in March and April.

? minutes

March

April 378 minutes

? minutes

58 minutes

```
  378 minutes
+  58 minutes
  ¹ ¹
  436 minutes
```

I can use the standard algorithm to add 378 minutes and 58 minutes. 436 minutes is the amount of time Matthew reads in March.

```
  436 minutes
+ 378 minutes
  ¹ ¹
  814 minutes
```

I can use the standard algorithm to add the time Matthew reads in March, 436 minutes, and the time he reads in April, 378 minutes, to find the total time he spends reading for both months.

Matthew read for 814 minutes in March and April.

Name _____ Date _____

1. Find the sums below.

   a. 47 m + 8 m

   b. 47 m + 38 m

   c. 147 m + 383 m

   d. 63 mL + 9 mL

   e. 463 mL + 79 mL

   f. 463 mL + 179 mL

   g. 368 kg + 263 kg

   h. 508 kg + 293 kg

   i. 103 kg + 799 kg

   j. 4 L 342 mL + 2 L 214 mL

   k. 3 kg 296 g + 5 kg 326 g

2. Mrs. Haley roasts a turkey for 55 minutes. She checks it and decides to roast it for an additional 46 minutes. Use a strip diagram to find the total minutes Mrs. Haley roasts the turkey.

3. A miniature horse weighs 268 fewer kilograms than a Shetland pony. Use the table to find the weight of a Shetland pony.

Types of Horses	Weight in kg
Shetland pony	_____ kg
American Saddlebred	478 kg
Clydesdale horse	_____ kg
Miniature horse	56 kg

4. A Clydesdale horse weighs as much as a Shetland pony and an American Saddlebred horse combined. How much does a Clydesdale horse weigh?

Lucy buys an apple that weighs 152 grams. She buys a banana that weighs 109 grams.

a. Estimate the total weight of the apple and banana by rounding.

   $152 \approx 200$
   $109 \approx 100$

   > I can round each number to the nearest hundred.

   $200 \text{ grams} + 100 \text{ grams} = 300 \text{ grams}$

   > I can add the rounded numbers to estimate the total weight of the apple and the banana. The total weight is about 300 grams.

b. Estimate the total weight of the apple and banana by rounding in a different way.

   $152 \approx 150$
   $109 \approx 110$

   > I can round each number to the nearest ten.

   $150 \text{ grams} + 110 \text{ grams} = 260 \text{ grams}$

   > I can add the rounded numbers to estimate the total weight of the apple and the banana. The total weight is about 260 grams.

c. Calculate the actual total weight of the apple and the banana. Which method of rounding was more precise? Why?

   152 grams
   $+109 \text{ grams}$
   $\overline{261 \text{ grams}}$

   Rounding to the nearest ten grams was more precise because when I rounded to the nearest ten grams, the estimate was 260 grams, and the actual answer is 261 grams. The estimate and the actual answer are only 1 gram apart! When I rounded to the nearest hundred grams, the estimate was 300 grams, which isn't that close to the actual answer.

   > I can use the standard algorithm to find the actual total weight of the apple and the banana.

Lesson 19: Estimate sums by rounding and apply to solve measurement word problems.

Name _____ Date _____

1. Cathy collects the following information about her dogs, Stella and Oliver.

Stella	
Time Spent Getting a Bath	Weight
36 minutes	32 kg

Oliver	
Time Spent Getting a Bath	Weight
25 minutes	7 kg

Use the information in the charts to answer the questions below.

a. Estimate the total weight of Stella and Oliver.

b. What is the actual total weight of Stella and Oliver?

c. Estimate the total amount of time Cathy spends giving her dogs a bath.

d. What is the actual total time Cathy spends giving her dogs a bath?

e. Explain how estimating helps you check the reasonableness of your answers.

Lesson 19: Estimate sums by rounding and apply to solve measurement word problems.

2. Dena reads for 361 minutes during Week 1 of her school's two-week long Read-A-Thon. She reads for 212 minutes during Week 2 of the Read-A-Thon.

   a. Estimate the total amount of time Dena reads during the Read-A-Thon by rounding.

   b. Estimate the total amount of time Dena reads during the Read-A-Thon by rounding in a different way.

   c. Calculate the actual number of minutes that Dena reads during the Read-A-Thon. Which method of rounding was more precise? Why?

A STORY OF UNITS – TEKS EDITION
Lesson 20 Homework Helper 3•2

1. Solve the subtraction problems below.

    a. 50 cm − 24 cm = **26 cm**

    > I can use mental math to solve this subtraction problem. I do not have to write it out vertically. I can also think of my work with quarters. I know $50 - 25 = 25$. But since I'm only subtracting 24, I need to add 1 more to 25. So, the answer is 26 cm.

    b. 507 g − 234 g

    $$\begin{array}{r} 507 \text{ g} \\ -\ 234 \text{ g} \\ \hline \end{array}$$

    > Before I subtract, I need to see if any tens or hundreds need to be unbundled. I can see that there are enough ones to subtract 4 ones from 7 ones. There is no need to unbundle a ten.

    $$\begin{array}{r} {}^{4\ 10}\!\!\!\not{5}\not{0}7 \text{ g} \\ -\ 234 \text{ g} \\ \hline \end{array}$$

    > But, I am still not ready to subtract. There are not enough tens to subtract 3 tens, so I need to unbundle 1 hundred to make 10 tens. Since I unbundled 1 hundred, there are now 4 hundreds left.

    $$\begin{array}{r} {}^{4\ 10}\!\!\!\not{5}\not{0}7 \text{ g} \\ -\ 234 \text{ g} \\ \hline 273 \text{ g} \end{array}$$

    > After unbundling, I see that there are 4 hundreds, 10 tens, and 7 ones. Now I am ready to subtract. Since I've prepared my numbers all at once, I can subtract left to right, or right to left. The answer is 273 grams.

Lesson 20: Decompose once to subtract measurements including three-digit minuends with zeros in the tens or ones place.

2. Renee buys 607 grams of cherries at the market on Monday. On Wednesday, she buys 345 grams of cherries. How many more grams of cherries did Renee buy on Monday than on Wednesday?

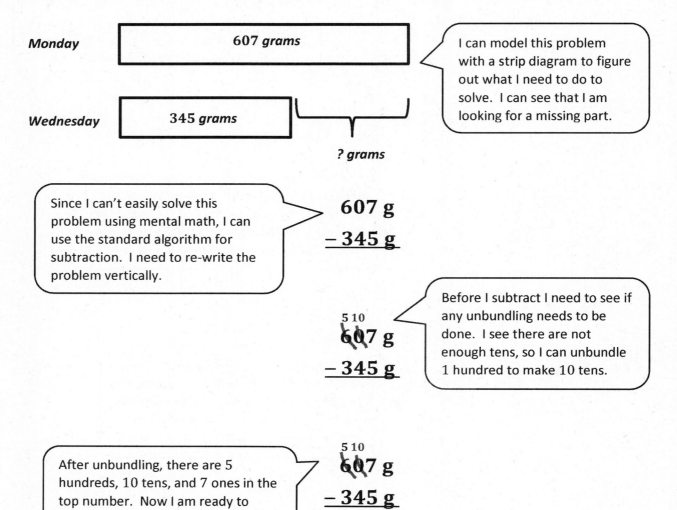

Renee buys 262 more grams of cherries on Monday than on Wednesday.

Name _____ Date _____

1. Solve the subtraction problems below.

   a. 70 L – 46 L

   b. 370 L – 46 L

   c. 370 L – 146 L

   d. 607 cm – 32 cm

   e. 592 cm – 258 cm

   f. 918 cm – 553 cm

   g. 763 g – 82 g

   h. 803 g – 542 g

   i. 572 km – 266 km

   j. 837 km – 645 km

2. The magazine weighs 280 grams less than the newspaper. The weight of the newspaper is shown below. How much does the magazine weigh? Use a strip diagram to model your thinking.

454 g

3. The chart to the right shows how long it takes to attend 3 events.

   a. Francesca's class play is 22 minutes shorter than Lucas's cooking competition. How long is Francesca's class play?

Lucas's cooking competition	180 minutes
Joey's dance recital	139 minutes
Francesca's class play	? minutes

   b. How much longer is Francesca's class play than Joey's dance recital?

1. Solve the subtraction problems below.

   a. 370 cm − 90 cm = **280 cm**

   > I can use mental math to solve this subtraction problem. I do not have to write it out vertically. Using the compensation strategy, I can add 10 to both numbers and think of the problem as 380 − 100, which is an easy calculation. The answer is 280 cm.

   b. 800 mL − 126 mL

   $$\begin{array}{r} \overset{7\ 10}{\cancel{8}\cancel{0}0} \text{ mL} \\ -\ 126 \text{ mL} \\ \hline \end{array}$$

   > Before I subtract, I need to see if any tens or hundreds need to be unbundled. There are not enough ones to subtract, so I can unbundle 1 ten to make 10 ones. But there are 0 tens, so I can unbundle 1 hundred to make 10 tens. Then there are 7 hundreds and 10 tens.

   $$\begin{array}{r} \overset{\ \ \ 9}{\overset{7\ \cancel{10}\ 10}{\cancel{8}\cancel{0}\cancel{0}}} \text{ mL} \\ -\ 126 \text{ mL} \\ \hline \end{array}$$

   > I still am not ready to subtract because I have to unbundle 1 ten to make 10 ones. Then there are 9 tens and 10 ones.

   $$\begin{array}{r} \overset{\ \ \ 9}{\overset{7\ \cancel{10}\ 10}{\cancel{8}\cancel{0}\cancel{0}}} \text{ mL} \\ -\ 126 \text{ mL} \\ \hline 674 \text{ mL} \end{array}$$

   > After unbundling, I see that I have 7 hundreds, 9 tens, and 10 ones. Now I am ready to subtract. Since I've prepared my numbers all at once, I can choose to subtract left to right, or right to left. The answer is 674 mL.

Lesson 21: Decompose twice to subtract measurements including three-digit minuends with zeros in the tens and ones places.

2. Kenny is driving from Los Angeles to San Diego. The total distance is about 175 kilometers. He has 86 kilometers left to drive. How many kilometers has he driven so far?

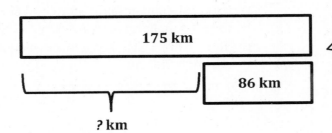

I can model this problem with a strip diagram to figure out what I need to do to solve. I can see that I am looking for a missing part.

Since I can't easily solve this problem using mental math, I can use the standard algorithm for subtraction. I can re-write the problem vertically.

$$\begin{array}{r} 175 \text{ km} \\ -86 \text{ km} \\ \hline \end{array}$$

$$\begin{array}{r} {}^{017} \\ \cancel{1}\cancel{7}5 \text{ km} \\ -86 \text{ km} \\ \hline \end{array}$$

Before I subtract, I need to see if any unbundling needs to be done. I can see there are not enough tens or ones, so I can unbundle 1 hundred to make 10 tens. After unbundling, there are 0 hundreds and 17 tens.

$$\begin{array}{r} {}^{16} \\ {}^{0\cancel{17}15} \\ \cancel{1}\cancel{7}\cancel{5} \text{ km} \\ -86 \text{ km} \\ \hline 89 \text{ km} \end{array}$$

I can unbundle 1 ten to make 10 ones. After unbundling, there are 0 hundreds, 16 tens, and 15 ones. I am ready to subtract. The answer is 89 kilometers.

Kenny has driven 89 km so far.

Name _____   Date _____

1. Solve the subtraction problems below.

   a. 280 g – 90 g

   b. 450 g – 284 g

   c. 423 cm – 136 cm

   d. 567 cm – 246 cm

   e. 900 g – 58 g

   f. 900 g – 358 g

   g. 4 L 710 mL – 2 L 690 mL

   h. 8 L 830 mL – 4 L 378 mL

2. The total weight of a giraffe and her calf is 904 kilograms. How much does the calf weigh? Use a strip diagram to model your thinking.

Giraffe
829 kg

Calf
? kg

3. It is 561 kilometers from San Antonio to Odessa. Salvador travels by car from San Antonio. He must travel 373 kilometers more before he reaches Odessa. How many kilometers has he traveled so far?

4. Mr. Nguyen fills two inflatable pools. The kiddie pool holds 185 liters of water. The larger pool holds 600 liters of water. How much more water does the larger pool hold than the kiddie pool?

Esther measures rope. She measures a total of 548 centimeters of rope and cuts it into two pieces. The first piece is 152 centimeters long. How long is the second piece of rope?

a. Estimate the length of the second piece of rope by rounding.

548 cm ≈ 500 cm
152 cm ≈ 200 cm

> I can round each number to the nearest hundred for my first estimate. I notice that both numbers are far from the hundred.

500 cm − 200 cm = 300 cm

The second piece of rope is about 300 cm long.

b. Estimate the length of the second piece of ribbon by rounding in a different way.

548 cm ≈ 550 cm
152 cm ≈ 150 cm

> I can round each number to the nearest ten for my second estimate. Wow, both numbers are close to the fifty! This makes it easy to calculate.

550 cm − 150 cm = 400 cm

The second piece of rope is about 400 cm long.

c. Precisely how long is the second piece of rope?

```
    4 14
   5̷4̷8 cm
 − 152 cm
   396 cm
```

> Before I am ready to subtract, I can unbundle 1 hundred for 10 tens.

The second piece of rope is precisely 396 cm long.

d. Is your answer reasonable? Which estimate was closer to the exact answer?

Rounding to the nearest ten was closer to the exact answer, and it was easy mental math. The estimate was only 4 cm away from the actual answer. So that's how I know my answer is reasonable.

> Comparing my actual answer with my estimate helps me check my calculation because if the answers are very different, I've probably made a mistake in my calculation.

Name _____ Date _____

Estimate, and then solve each problem.

1. Melissa and her mom go on a road trip. They drive 87 kilometers before lunch. They drive 59 kilometers after lunch.

    a. Estimate how many more kilometers they drive before lunch than after lunch by rounding to the nearest 10 kilometers.

    b. Precisely how much farther do they drive before lunch than after lunch?

    c. Compare your estimate from (a) to your answer from (b). Is your answer reasonable? Write a sentence to explain your thinking.

2. Amy measures rope. She measures a total of 393 centimeters of rope and cuts it into two pieces. The first piece is 184 centimeters long. How long is the second piece of rope?

    a. Estimate the length of the second piece of rope by rounding in two different ways.

    b. Precisely how long is the second piece of rope? Explain why one estimate was closer.

Lesson 22: Estimate differences by rounding and apply to solve measurement word problems.

3. The weight of a chicken leg, steak, and ham are shown to the right. The chicken and the steak together weigh 341 grams. How much does the ham weigh?

   989 grams

   a. Estimate the weight of the ham by rounding.

   b. How much does the ham actually weigh?

4. Kate uses 506 liters of water each week to water plants. She uses 252 liters to water the plants in the greenhouse. How much water does she use for the other plants?

   a. Estimate how much water Kate uses for the other plants by rounding.

   b. Estimate how much water Kate uses for the other plants by rounding a different way.

   c. How much water does Kate actually use for the other plants? Which estimate was closer? Explain why.

Mia measures the lengths of three pieces of wire. The lengths of the wires are recorded to the right.

Wire A	63 cm ≈ __60__ cm
Wire B	75 cm ≈ __80__ cm
Wire C	49 cm ≈ __50__ cm

a. Estimate the total length of Wire A and Wire C. Then, find the actual total length.

> I can round the lengths of all the wires to the nearest ten.

Estimate: 60 cm + 50 cm = 110 cm

> I can add the rounded lengths of Wires A and C to find an estimate of their total length.

Actual: 63 cm + 49 cm = 112 cm
          / \
        62   1   50

> I can use mental math to solve this problem. I do not have to write it out vertically. I can break apart 63 as 62 and 1. Then I can make the next ten to 50, and then add the 62.

The total length is 112 cm.

b. Subtract to estimate the difference between the total length of Wires A and C and the length of Wire B. Then, find the actual difference. Model the problem with a strip diagram.

Estimate: 110 cm − 80 cm = 30 cm

Actual: 112 cm − 75 cm = 37 cm

| Wire A + Wire C | 112 cm |
| Wire B | 75 cm | ? cm |

> From the strip diagram, I see that I need to solve for an unknown part.

```
   10 12
   1̶1̶2̶ cm
 −  75 cm
    37 cm
```

The difference is 37 cm.

> I can write this problem vertically. I can unbundle 1 ten for 10 ones. I can rename 112 as 10 tens and 12 ones. Then I am ready to subtract.

Name _____ Date _____

1. There are 153 milliliters of juice in 1 carton. A three-pack of juice boxes contains a total of 459 milliliters.

   a. Estimate, and then find the actual total amount of juice in 1 carton and in a three-pack of juice boxes.

      153 mL + 459 mL ≈ _____ + _____ = _____

      153 mL + 459 mL = _____

   b. Estimate, and then find the actual difference between the amount in 1 carton and in a three-pack of juice boxes.

      459 mL − 153 mL ≈ _____ − _____ = _____

      459 mL − 153 mL = _____

   c. Are your answers reasonable? Why?

2. Mr. Williams owns a gas station. He sells 367 liters of gas in the morning, 300 liters of gas in the afternoon, and 219 liters of gas in the evening.

   a. Estimate, and then find the actual total amount of gas he sells in one day.

   b. Estimate, and then find the actual difference between the amount of gas Mr. Williams sells in the morning and the amount he sells in the evening.

3. The Blue Team runs a relay. The chart shows the time, in minutes, that each team member spends running.

   a. How many minutes does it take the Blue Team to run the relay?

Blue Team	Time in Minutes
Jen	5 minutes
Kristin	7 minutes
Lester	6 minutes
Evy	8 minutes
Total	

   b. It takes the Red Team 37 minutes to run the relay. Estimate, and then find the actual difference in time between the two teams.

4. The lengths of three banners are shown to the right.

   a. Estimate, and then find the actual total length of Banner A and Banner C.

Banner A	437 cm
Banner B	457 cm
Banner C	332 cm

   b. Estimate, and then find the actual difference in length between Banner B and the combined length of Banner A and Banner C. Model the problem with a strip diagram.

Credits

Great Minds® has made every effort to obtain permission for the reprinting of all copyrighted material. If any owner of copyrighted material is not acknowledged herein, please contact Great Minds for proper acknowledgment in all future editions and reprints of these modules.